ANALOG CIRCUITS AND SIGNAL PROCESSING

Series Editors
Mohammed Ismail, The Ohio State University
Mohamad Sawan, École Polytechnique de Montréal

T0189215

For further volumes:
http://www.springer.com/series/7381

Yongjian Tang · Hans Hegt
Arthur van Roermund

Dynamic-Mismatch Mapping for Digitally-Assisted DACs

 Springer

Yongjian Tang
MaxLinear Inc.
Carlsbad, CA, USA

Arthur van Roermund
Department of Electrical Engineering
Eindhoven University of Technology
Eindhoven, The Netherlands

Hans Hegt
Department of Electrical Engineering
Eindhoven University of Technology
Eindhoven, The Netherlands

ISBN 978-1-4899-9291-8   ISBN 978-1-4614-1250-2 (eBook)
DOI 10.1007/978-1-4614-1250-2
Springer New York Heidelberg Dordrecht London

Printed on acid-free paper

Springer is part of Springer Science+Business Media (www.springer.com)

Contents

1   **Introduction**   1

2   **Digital-to-Analog Converters**   5
    2.1   Introduction to DACs   5
        2.1.1   Time Domain Response   5
        2.1.2   Frequency Domain Response   7
        2.1.3   Applications   8
    2.2   Performance Specifications   8
        2.2.1   Static Performance (DC) Specifications   8
        2.2.2   Dynamic Performance (AC) Specifications   10
    2.3   Architectures   12
        2.3.1   Binary Architecture   12
        2.3.2   Thermometer (Unary) Architecture   13
        2.3.3   Segmented Architecture   14
    2.4   Physical Implementations   14
        2.4.1   Resistor DAC   14
        2.4.2   Capacitor DAC   15
        2.4.3   Current-Steering DAC (CS-DAC)   16
    2.5   State of The Art   17
    2.6   Conclusions   21

3   **Modeling and Analysis of Performance Limitations in CS-DACs**   23
    3.1   Static Mismatch Error   23
        3.1.1   Error Source: Amplitude Error   24
        3.1.2   Effect on Static Performance   26
        3.1.3   Effect on Dynamic Performance   26
    3.2   Dynamic Mismatch Error   32
        3.2.1   Error Sources: Amplitude and Timing Errors   32
        3.2.2   Dynamic Mismatch in Frequency Domain   45
        3.2.3   New Parameters to Evaluate Dynamic Matching:
                Dynamic-DNL and Dynamic-INL   47
        3.2.4   Comparison to Traditional Static DNL & INL   50

3.3   Non-mismatch Error   51
      3.3.1   Sampling Jitter   51
      3.3.2   Common Duty-Cycle Error   56
      3.3.3   Finite Output Impedance   61
      3.3.4   Data-Dependent Switching Interference......................   64
3.4   Summary of Performance Limitations   66
3.5   Conclusions   68

4   Design Techniques for High-Performance Intrinsic and
    Smart CS-DACs   69
    4.1   Introduction to Smart DACs   69
    4.2   Design Techniques for Intrinsic DACs   71
          4.2.1   Non-mismatch-Error Focused Techniques   71
          4.2.2   Mismatch-Error Focused Techniques   77
    4.3   Design Techniques for Smart DACs   79
          4.3.1   Analog Calibration Techniques   80
          4.3.2   Digital Calibration Techniques...............................   81
    4.4   Summary of Design Techniques for Intrinsic and Smart DACs   86
    4.5   Conclusions   89

5   A Novel Digital Calibration Technique: Dynamic-Mismatch
    Mapping (DMM)   91
    5.1   Theory of Dynamic-Mismatch Mapping.............................   91
    5.2   Measurement of Dynamic-Mismatch Error..........................   95
          5.2.1   Measurement Flow   95
          5.2.2   Sine-Wave Demodulation *vs.* Square-Wave Demodulation...   100
          5.2.3   Weight Function Between Amplitude and Timing Errors   103
    5.3   Theoretical Evaluation of DMM   105
          5.3.1   Effect of f_m on Performance Improvement   105
          5.3.2   Robustness of DMM   113
          5.3.3   Application of DMM and Comparison
                  to Other Techniques   113
    5.4   Conclusions   116

6   An On-chip Dynamic-Mismatch Sensor Based on a Zero-IF Receiver   119
    6.1   Architecture Considerations   119
    6.2   Analog Front-End Design   121
          6.2.1   Circuit Design...   121
          6.2.2   Signal Transfer Function   125
          6.2.3   Noise Analysis   128
    6.3   ADC Design...   134
    6.4   Overall Performance   134
    6.5   Conclusions   135

7 Design Example 137
   7.1 Overview 137
   7.2 A 14-bit 650 MS/s Intrinsic DAC Core 138
      7.2.1 Circuit Design.. 138
      7.2.2 Experimental Results 140
      7.2.3 Comparison to Other Works.................................. 142
   7.3 A 14-bit 200 MS/s Smart DAC with DMM 143
      7.3.1 Circuit Design.. 143
      7.3.2 Experimental Results 145
      7.3.3 Benchmark 149
   7.4 Conclusions 155

8 Conclusions 157

References.. 159

Index 163

Symbols and Abbreviations

Symbol	Description	Unit
ADC	Analog-to-digital converter	
CML	Current-mode logic	
D_{com}	Common duty-cycle error	seconds
DAC	Digital-to-analog converter	
DEM	Dynamic element matching	
DMM	Dynamic-mismatch mapping	
DNL	Differential non-linearity	LSB
Dynamic-DNL	Dynamic differential non-linearity	LSB
Dynamic-INL	Dynamic integral non-linearity	LSB
DSP	Digital signal processor	
ENOB	Effective number of bits	bit
f_i	Input signal frequency	Hz
f_m	Modulation or measurement frequency	Hz
f_s	Sampling frequency	Hz
IM3	Third-order intermodulation	dBc
IMD	Intermodulation distortion	dBc
INL	Integral non-linearity	LSB
LO	Local oscillator	
LVDS	Low-voltage differential signaling	
NRZ	Non-return-to-zero	
NSD	Noise power spectral density	dBm/Hz
OTA	Operational transconductance amplifier	
RZ	Return-to-zero	
SFDR	Spurious-free dynamic range	dB
SMM	Static-mismatch mapping	
SNR	Signal-to-noise ratio	dB
SNDR	Signal-to-(noise+distorion) ratio	dB
TIA	Trans-impedance amplifier	
THD	Total harmonic distortion	dBc

x

Symbol	Description	Unit
T_s	Sampling period	seconds
ZOH	Zero-order-hold	
Z_{out}	Output impedance	Ω
σ_{amp}	Deviation of Gaussian distributed amplitude errors	%
σ_{timing}	Deviation of Gaussian distributed timing errors	seconds
σ_{jitter}	Deviation of Gaussian distributed jitter	seconds

Chapter 1
Introduction

Since the invention of the first semiconductor transistor in the 1940s and the breakthrough in the 1960s, microelectronics has been one of the most rapidly developed technologies in the past few decades. The advanced microelectronics techniques, such as integrated circuits (ICs), dramatically reformed our daily life and scientific research, such as space technique, sensing technique, telecommunications, computer science and multimedia entertainment.

As technology is moving to deep sub-micron or even nanometer scale, the complementary metal-oxide-semiconductor (CMOS) technology has become the dominant manufacturing technology for microelectronics in very large scale integration (VLSI) applications. Digital integrated circuits directly benefit from this CMOS technology scaling, since the minimal gate length of the transistor has a scaling factor of 0.7 from generation to generation (e.g. $0.18\,\mu m \rightarrow 0.13\,\mu m \rightarrow 90\,nm \rightarrow 65\,nm$). This scaling to ever smaller dimensions leads to higher transistor-integration density, faster circuit speed, lower power dissipation and significantly reduced cost per function. As a result, nowadays, more and more signal processing is preferred to be performed in the digital domain by digital signal processors (DSPs). This trend significantly increases the demand for high quality interface circuits between analog and digital domain. Data converters, i.e. Analog-to-Digital converters (ADCs) and Digital-to-Analog converters (DACs), as essential devices in interface circuits, are required to achieved high performance with increased signal and sampling frequencies. In many emerging applications, such as wideband or software-defined multi-mode communications, ADCs and DACs are already one of the major performance bottlenecks of the whole system. The research on high-speed high-performance data converters has become one of the key topics in microelectronics, in both academia and industry.

The aim of this book is to develop design techniques for high-speed high-performance smart DACs, especially designing a DAC with high dynamic performance, e.g. high linearity, is the main concern of this work. The work focuses on Nyquist DACs with current-steering architecture since that is the most suitable topology for high speed applications. For investigating fundamental performance

Y. Tang et al., *Dynamic-Mismatch Mapping for Digitally-Assisted DACs*, Analog Circuits and Signal Processing 92, DOI 10.1007/978-1-4614-1250-2_1,
© Springer Science+Business Media New York 2013

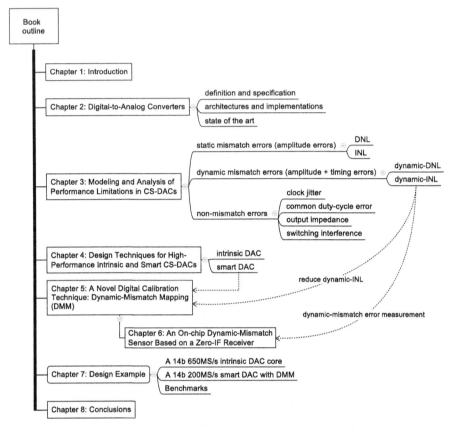

Fig. 1.1 Book outline

limitations, the effect of various error sources need to be analyzed. Based on that outcome, smart design techniques can be developed to overcome technology limitations so that a high performance can be achieved.

Figure 1.1 shows the outline of this book. Chapter 2 covers the basics of Nyquist DACs, such as the definition, performance specifications, architectures and physical implementations. Recently published state-of-the-art Nyquist DACs are also summarized in that chapter.

In Chap. 3, mismatch and non-mismatch errors are analyzed to investigate their influence on the performance of current-steering DACs. In the signal frequency range from DC to a few hundreds of MHz, mismatch errors, such as amplitude and timing errors, are typically the dominant error sources in the linearity of a DAC. As signal and sampling frequencies increase, the effect of timing errors becomes more and more dominant than that of amplitude errors. Traditional integral-nonlinearity (INL) and differential-nonlinearity (DNL) are based on the static matching behavior between current cells, i.e. only based on amplitude errors. In Chap. 3, two new parameters, named dynamic-INL and dynamic-DNL, are introduced to evaluate the dynamic matching behavior between current cells. Compared to traditional static INL and DNL, dynamic-INL and dynamic-DNL include both amplitude and timing errors, resulting in a new methodology to improve the performance of DACs.

Chapter 4 introduces the concept of smart DACs. A smart DAC is an intrinsic DAC with additional techniques to acquire actual chip information and improve the performance, yield, reliability or flexibility. Existing design techniques for high-performance intrinsic and smart DACs are categorized and discussed.

Based on the concept of the dynamic-INL, Chap. 5 introduces a novel digital calibration technique, called dynamic-mismatch mapping (DMM), to correct the effect of both amplitude and timing errors in a digital way. Theoretical proofs of the proposed DMM technique are given with dedicated explanations. The application of the DMM technique and the comparison to other calibration techniques are also discussed in Chap. 5. Since the proposed DMM technique requires the dynamic-mismatch errors to be accurately measured, an on-chip dynamic-mismatch sensor is designed in Chap. 6. In order to verify the proposed DMM technique, Chap. 7 gives a design example of a 14-bit current-steering DAC. The silicon experimental results of a 14-bit 650MS/s intrinsic DAC core and a 14-bit 200MS/s smart DAC with DMM are demonstrated. Benchmark comparison shows that this design achieves state-of-the-art performance.

Finally, conclusions are drawn in Chap. 8.

Chapter 2
Digital-to-Analog Converters

In this chapter, the concept and performance specifications of digital-to-analog converters (DACs) are reviewed. Different DAC architectures and physical implementations are introduced. Recently published state-of-the-art DACs are summarized to show the performance limitations.

2.1 Introduction to DACs

In this section, the general function and applications of digital-to-analog converters (DACs) are briefly discussed.

2.1.1 Time Domain Response

In electronics, a Digital-to-Analog Converter (DAC) is a device that converts a finite-precision digital-format number (the input, typically a finite-length binary-format number) to an analog electrical quantity (such as voltage, current or electric charge). Nowadays, with the development of digital technologies, for easy storage and processing, most analog signals are digitized by Analog-to-Digital converters (ADCs) and are processed by digital signal processors (DSPs) [1]. However, our perceptual world is still analog so that the digital signal has to be converted back into analog domain such that, for example, the data can be transmitted with high signal quality in communication systems or a human being can hear a music or watch a video. Therefore, a DAC is an essential device in scientific research, industry control and people's daily life.

Figure 2.1 shows a simplified signal chain with a DAC in a wireless transceiver. The input information to a DAC can come from two sources: the digital signal processor (DSP) or the Analog-to-Digital converter (ADC). The difference between

Y. Tang et al., *Dynamic-Mismatch Mapping for Digitally-Assisted DACs*, Analog Circuits and Signal Processing 92, DOI 10.1007/978-1-4614-1250-2_2,

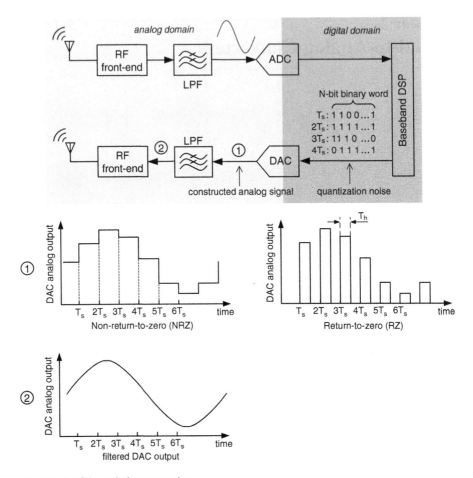

Fig. 2.1 DAC in a wireless transceiver

these two sources is that the information from the ADC is generated by digitizing an analog signal, while the DSP may directly generate this information. In order to construct an analog signal, there are two basic types of DAC output format: non-return-to-zero (NRZ) and return-to-zero (RZ). As shown in Fig. 2.1, for NRZ, the DAC updates its analog output according to its digital input at a fixed time interval of T_s and holds the output, where T_s is called updating or sampling period. For RZ, after updating the output at each time interval T_s, the DAC holds the output only for a certain time (T_h), then goes back to zero. In both cases, the DAC's output is held for a certain time T_h, where $0 < T_h \leq T_s$, known as zero-order-hold (ZOH). Compared to NRZ DACs, RZ DACs have a lower output power due to return-to-zero, e.g. half power when $T_h = 0.5T_s$. The output of a DAC is typically a stepwise or pulsed analog signal and can be low-pass filtered to construct the required analog signal.

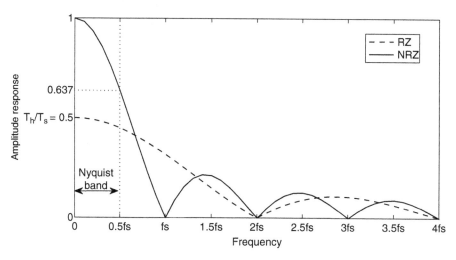

Fig. 2.2 Magnitude of frequency responses of ideal NRZ and RZ DACs

Form a simple point of view, the DAC performs the reverse operation of the ADC. However, it should be noticed that unlike the ADC, the DAC itself does not add any quantization noise because the quantization noise is already generated before the DAC. This is due to the finite quantization levels of the ADC or the finite word length of the DSP, i.e. finite precision. The word length of the binary digital input of the DAC, i.e. N-bit, is called the number of bits. Though the DAC does not generate quantization noise, it will most likely generate conversion errors due to the non-ideality of the DAC. The conversion errors are often input-signal related and generate harmonic distortion. The relation between the conversion errors and the performance of DACs will be discussed in Chap. 3.

2.1.2 Frequency Domain Response

The magnitude of the frequency responses of ideal NRZ and RZ DACs are shown in Fig. 2.2, where the RZ DAC example has a holding time (T_h) of a half of the sampling period (T_s). f_s ($= \frac{1}{T_s}$) is called updating rate or sampling frequency. The frequency responses are sinc-shaped because of the zero-order hold (ZOH) function in the DAC's output, and the shape is dependent on $\frac{T_h}{T_s}$. As seen, in the Nyquist band, the RZ DAC has a larger signal attenuation than the NRZ DAC. Compared to the NRZ DAC, the maximum signal loss for the RZ DAC is $|20log_{10}\frac{T_h}{T_s}|$dB at DC, e.g. 6 dB power loss at DC for $\frac{T_h}{T_s} = 0.5$. However, in the Nyquist band, a RZ DAC has a more flat magnitude response than a NRZ DAC. In this example, the magnitude drop is 3.9 dB for NRZ and 0.9 dB for RZ at 0.5 f_s, respectively. For some applications where a flat magnitude response is required in the Nyquist band, an anti-sinc digital filter can be placed before the DAC to compensate the sinc attenuation.

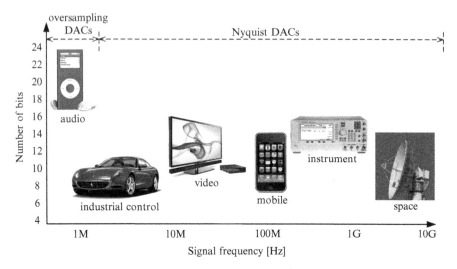

Fig. 2.3 Application examples of digital-to-analog converters

2.1.3 Applications

Figure 2.3 shows a few typical applications of DACs. Oversampling DACs are dominant in audio applications, where 16- to 24-bit is required in a kHz signal frequency range. This work focuses on Nyquist DACs for high signal frequencies (MHz- to GHz-range). This kind of DAC is widely used in high-speed instruments and telecommunications.

2.2 Performance Specifications

In this section, static and dynamic performance specifications of DACs are briefly introduced.

2.2.1 Static Performance (DC) Specifications

Static performance specifications introduced below are used to evaluate the DC performance of DACs.

2.2.1.1 Offset and Gain Errors

The offset error of a DAC is defined as the deviation of the linearized transfer curve of the DAC output from the ideal zero. The linearized transfer curve is based on the

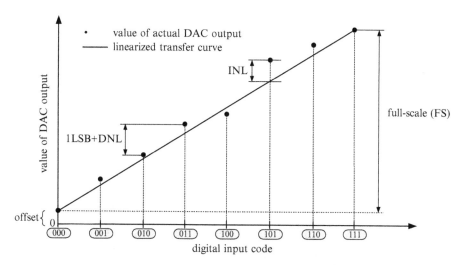

Fig. 2.4 DC specs of a DAC

actual DAC output, either a simple min–max line connecting the minimal and the maximal DAC output value or a best-fit line of all the output values of the DAC. A 3-bit DAC example is shown in Fig. 2.4 with a simple line as the linearized transfer curve. The difference between the minimal value and the maximum value of the linearized transfer curve is called full-scale (FS) output range. The error between 1 and the ratio of the actual full-scale range over the ideal full-scale range is called the gain error (in percentage). The offset can be easily compensated by a DC auxiliary DAC and the gain error can be corrected by adjusting the full-scale range settings. Since the offset and gain errors do not introduce non-linearity, they have no effect on the spectral performance of DACs.

2.2.1.2   Integral Non-linearity (INL)

As shown in Fig. 2.4, integral non-linearity (INL) is defined as the deviation of the actual DAC output from the linearized transfer curve at every code input. The INL_{max} is the worst value of the INL, as shown in (2.1), where N is the number of bits of the DAC. As seen, the INL directly reflects the static linearity of the DAC.

$$INL(code) = out_{dac}(code) - (\text{offset} + 1\text{LSB\_stepsize} \times code),$$

$$\text{where } 1\text{LSB\_stepsize} = \frac{\text{full-scale DAC output}}{2^N - 1}$$

$$INL_{max} = max(INL(code)), \text{code=0}\sim\text{full-scale digital input code} \quad (2.1)$$

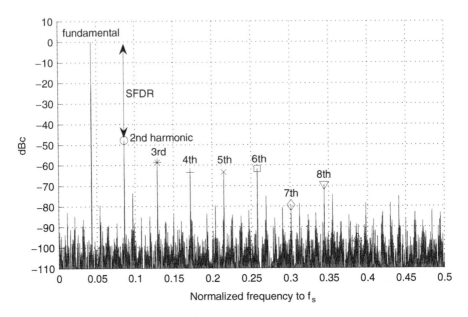

Fig. 2.5 An example of DAC output spectrum

2.2.1.3 Differential Non-linearity (DNL)

As shown in Fig. 2.4, the differential non-linearity (DNL) is the deviation of the actual step size from the ideal step size (1LSB) between any two adjacent digital input codes. The DNL_{max} is the worst case of the DNL.

$$DNL(code) = out_{dac}(code) - out_{dac}(code - 1) - 1\text{LSB\_stepsize}$$

$$DNL_{max} = max(DNL(code)), \text{code=1}\sim\text{full-scale digital input code} \quad (2.2)$$

2.2.2 Dynamic Performance (AC) Specifications

Dynamic performance specifications are used to evaluate the AC performance of DACs. These parameters are very important in many applications, such as in high-speed communication systems which is one of the targeted applications of this work.

2.2.2.1 Single-Tone SFDR/THD/NSD/SNR/SNDR

Figure 2.5 shows an example of the output spectrum of a Nyquist DAC with a single-tone sine-wave input. The frequency axis is normalized to the sampling frequency (f_s). Several parameters are defined in the frequency domain to evaluate the dynamic performance of the DAC.

Spurious-free Dynamic Range (SFDR):

The ratio, in decibels, between the power of the fundamental component of the constructed output sine wave and the power of the largest spurious tone observed (excluding the DC component) in the frequency domain. Typically a high SFDR is required to suppress spurious emissions, especially in communication systems.

Total Harmonic Distortion (THD):

The total power of all harmonics of the reconstructed output sine wave. The THD can be expressed in decibels if it is relative to the power of the fundamental component of the constructed output sine wave.

Noise Power Spectral Density (NSD):

The power density of the noise at the DAC's output in the frequency domain. It can be specified in dBm/Hz.

Signal-to-Noise Ratio (SNR):

The ratio of the power of the measured output signal to the integrated power of the noise floor in the Nyquist band ($[0, \frac{\text{sampling frequency}}{2}]$, except DC and harmonics). The value for SNR is expressed in decibels.

Signal-to-(Noise+Distorion) Ratio (SNDR):

The ratio of the power of the measured output signal to the integrated power of the noise floor in the Nyquist band plus the total power of the harmonics. The SNDR directly relates to the SNR and THD.

2.2.2.2   Two-Tone Intermodulation Distortion (IMD)

When a two-tone signal is applied to a nonlinear system, intermodulation distortion products are generated. Assuming the frequencies of the two tones are f_1 and f_2 ($f_1 < f_2$), the spectral components which are most close to the fundamental output tones are two third-order intermodulation distortion components $2f_1 - f_2$ (IM3_{left}) and $2f_2 - f_1$ (IM3_{right}), as shown in Fig. 2.6. Then, the IM3 is defined as the worse one between IM3_{right} and IM3_{left}. As seen, if the frequencies of the two input tones are adjacent with close spacing, the IM3 falls very close to the desired signals. This is strongly not desired since the IM3 is then very difficult to be filtered out.

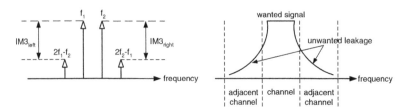

Fig. 2.6 Graphic representation of IM3 and ACLR

The adjacent channel leakage ratio (ACLR) is also used to indicate the intermodulation performance, especially in multi-channel broadband systems such as WCDMA, CDMA2000, WiMAX, LTE, etc.. It is defined as a ratio, in dBc, of the transmitted power within a desired channel to the power in its adjacent channel. It has the same generation mechanism as the intermodulation and can be related to the IMDs.

2.3 Architectures

According to different decoding schemes, DACs have three basic architectures: binary, thermometer and segmented architectures. There are also other types of architectures which are optimized for specific input signals, such as sine-weighted DACs. In this section, only basic DAC architectures for general input signals are discussed.

2.3.1 Binary Architecture

Since the input to a DAC is typically a binary digital word, the most straightforward way to implement the function of the DAC is to let every input bit corresponds to a binary-weighted element (voltage, current or charge). An example of 4-bit binary-coded DAC is shown in Fig. 2.7.

The advantage of a binary-coded DAC is that its decoding circuit and the number of switches are minimal, i.e. its chip area and power consumption are small. The disadvantage is that the ratio between the least significant element and the most significant element is so large that the matching between them is difficult to be guaranteed, resulting in large DNL and INL errors. Another drawback is that if the switching of the elements are not perfectly synchronized, large glitch errors occur during input code transitions, especially when the most significant element is being switched.

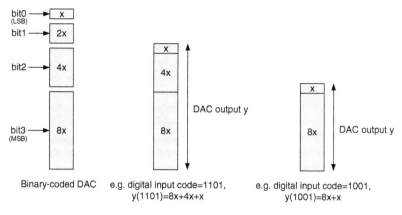

Fig. 2.7 A 4-bit binary-coded DAC example

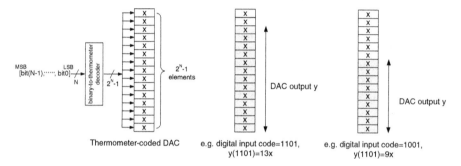

Fig. 2.8 A 4-bit thermometer-coded DAC example

2.3.2   Thermometer (Unary) Architecture

In order to overcome the drawbacks of the binary-coded DAC architecture, a thermometer-coded DAC architecture has been developed. As shown in Fig. 2.8, an N-bit thermometer-coded DAC has $2^N - 1$ unary elements. Those unary elements are switched on or off in a certain sequence according to the input digital code. Compared to the binary-coded architecture, the thermometer-coded architecture reduces the INL/DNL and glitch errors. The costs are: a binary-to-thermometer decoder is needed and lots of switches have to be synchronized. Since the area and power consumption of the decoder and switches are exponentially increasing with the number of bits, a full thermometer-coded DAC architecture is seldom used with N above 10-bit.

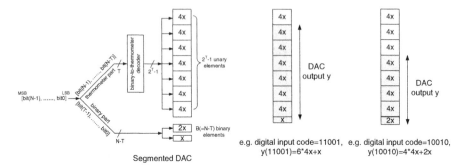

Fig. 2.9 A 5-bit 3T-2B segmented DAC example

2.3.3 Segmented Architecture

The segmented architecture is the most widely used DAC architecture since it balances the pros and cons of binary and thermometer architectures. For a segmented DAC, part of the input digital code, typically several most significant bits, are implemented as unary elements and the other part is implemented as binary elements. In the 5-bit DAC example shown in Fig. 2.9, the first three bits are implemented as a thermometer-coded sub-DAC and the last two bits are implemented as a binary-coded sub-DAC. As a result, the DAC has a 3thermometer-2binary (3T-2B) segmented architecture. How to segment the total bits into thermometer and binary parts is a trade-off between performance, area and power consumption. In a segmented DAC, the thermometer part is typically dominant in the whole performance of the DAC.

2.4 Physical Implementations

Depending on how an element is implemented, there are three basic DAC physical implementations: resistor DACs, capacitor DACs and current-steering DACs. In the following sections, examples of basic implementations of these three types of DACs and their applications are discussed.

2.4.1 Resistor DAC

Figure 2.10 shows a frequently used R-2R ladder DAC. By connecting or disconnecting the resistors, the output voltage (Vout) is controlled by the input binary bits. The DAC accuracy depends on the matching of the resistors. Speed and linearity are main limits of resistor type DACs due to the nonlinear resistors and the bandwidth and linearity of the output buffer.

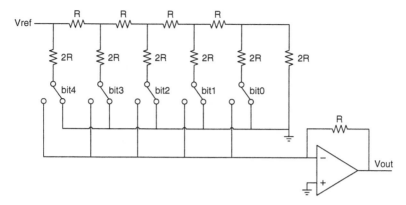

Fig. 2.10 A 5-bit R-2R ladder DAC example

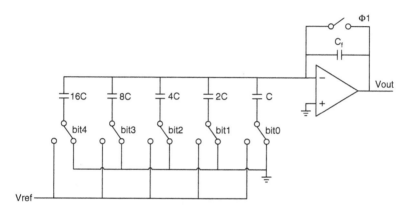

Fig. 2.11 A 5-bit switched-cap DAC example

2.4.2 *Capacitor DAC*

Figure 2.11 shows an example of a switched-capacitor DAC. The operation needs two phases. During phase $\phi 1$, the input capacitors are connected either to a reference voltage (Vref) or to ground according to the input digital code, and the feedback capacitor is shorted. During phase $\phi 2$, all input capacitors are switched to ground and the feedback capacitor is connected around the amplifier. Based on charge conservation, the output voltage (Vout) is a fraction of Vref which is set by the input digital code. Similar to the resistor DAC, the capacitor DAC's accuracy depends on the matching of the capacitors. Speed and linearity are also main limits of this type of DAC. The advantage of capacitor DACs is that the power consumption is quite low since only a certain charge needs to be transferred.

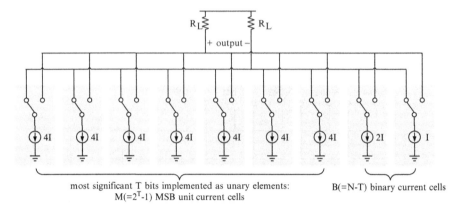

most significant T bits implemented as unary elements: B(=N-T) binary current cells
M(=2^T-1) MSB unit current cells

Fig. 2.12 A 5-bit 3T-2B segmented current-steering DAC example

2.4.3 Current-Steering DAC (CS-DAC)

With the rapid development of communication systems, such as Direct-Digital-Synthesis (DDS) and novel RF transceivers in new applications, high-speed and high-resolution DACs are required. In these applications, very high sampling-rate DACs, which often need to be operated at hundreds of MHz and drive a 50 ohm load, can directly generate RF/IF signals. Consequently, it is unnecessary to use traditional mixers for up-conversion. This is very suitable for multi-standard or long-term evolution applications because in this way, most of the signal processing can be done in the digital domain. The current-steering DAC is a suitable architecture for such applications, because of its intrinsic high speed and driving capability.

An example of a 5-bit 3T-2B segmented current-steering DAC architecture with a differential output is shown in Fig. 2.12. In this figure, N is the total number of bits of the DAC (for simplicity, the binary-to-thermometer decoder shown in Fig. 2.9 is not shown here). As seen, most significant T bits are implemented as unary elements, called MSB unit current cells which all provide the same current. The remaining B (=N-T) bits are implemented as binary elements, called binary current cells whose currents are binary-weighted. The current cell consists of a current source and differential switches: the current is switched to the positive output node or to the negative output node according to the input digital bits. Therefore, all current cells are acting as switched-current (SI) cells. The DAC's accuracy relies on the matching between current sources.

Since the output of a current-steering DAC is a current and has a high output impedance, it has very fast conversion speed and good intrinsic driving ability for low impedance loading. For high speed applications, a loading resistor (R_L, typically $25\Omega \sim 300\Omega$) converts the current output to a voltage. The differential output voltage swing is $2I_{FS}R_L$, where I_{FS} is the full-scale output current of the DAC. As seen, a larger loading resistor leads to a larger output voltage swing, i.e. larger delivered power. Because the DAC output is a current and the resistor performs a linear I–V conversion, in theory, the linearity of the DAC is only determined by the linearity of the output current. Therefore, the linearity of the DAC, such as the SFDR or IM3, is independent of the output swing, as long as the linearity of the output current is not compromised by the large voltage swing, e.g. if there is not enough voltage headroom for correct current source biasing. However, in practice, due to technology limitations, the linearity of the output current can be compromised by a larger output voltage swing, so does the linearity of the DAC.

2.5 State of The Art

Table 2.1 lists the main state-of-the-art Nyquist DACs published in the last twelve years. As seen, high speed, high performance and low power are major research trends. Especially, driven by new communication applications, the DAC is moving to the RF frequency where high speed and high dynamic performance are both required. How to meet those requirements is a challenge for the DAC design and will be addressed in this work.

Figure 2.13 shows the sampling frequency of the published DACs in Table 2.1 versus the process technology. As expected, due to higher f_T, a BiCMOS or bipolar technology can achieve a much higher sampling frequency than a CMOS technology. In the same category of CMOS technology, in general, more advanced technology nodes can achieve a higher sampling frequency. However, as seen, the sampling frequencies of most of the Nyquist CMOS DACs are still in the range of 100 MHz to 1 GHz. One reason for this is that in most traditional applications, due to a low signal frequency, a high-linearity performance is typically required rather than a very high sampling frequency. With increasing signal frequency in emerging applications, a DAC with >1 GHz sampling frequency and high-linearity performance becomes very attractive [2, 12].

The SFDR of these published DACs at very low signal frequencies (near DC) versus static effective number of bits (static ENOB, based on the INL) is summarized in Fig. 2.14. As seen, most of these DACs have an 11–15 bit ENOB for their static performance, which is limited by the static matching accuracy. The SFDRs at very low signal frequencies are mostly located between 70–85 dBc, which are mainly limited by the static linearity of DACs, i.e. the INLs.

The SFDR of these DACs at high input signal frequencies (near Nyquist frequency, i.e near half of sampling frequency, unless specified in Table 2.1) are plotted in Fig. 2.15. At tens or hundreds of MHz signal frequencies, not only static non-linearity but also dynamic non-idealities limit the DAC's dynamic performance. As seen, the SFDR at signal frequencies below 100 MHz is hardly higher than 80 dBc, and it drops pretty fast with further increasing signal frequencies. For CMOS technology, the state-of-the-art DACs achieve 60 dBc around 500 MHz

Table 2.1 State-of-the-art Nyquist DACs

Ref.	Year	Bits	INL/DNL [LSB]	fs [MS/s]	SFDR@low fi [dBc]	SFDR@high fi [dBc]	Technology	Power
[2]	ISSCC'09	12	0.5/0.3	2,900	74	60[a]	65 nm CMOS, 2.5 V	188 mW
[3]	ISSCC'07	13	0.8/0.4	200	83.7	54.5	0.13 um CMOS, 1.5 V	25 mW
[4]	ISSCC'06	14	–	100	74.4	77.8	0.18 um CMOS, 1.8 V	150 mW
[5]	ISSCC'06	6	–	20,000	–	50[b]	0.18 um SiGe, 1.8 V	360 mW
[6]	ISSCC'06	9	1/0.5	2	–	–	0.5 um CMOS, 5 V	0.3 mW
[7]	ISSCC'05	15	8	1,200	72	63	0.35 um BiCMOS, 3.3 V	6 W
[8]	ISSCC'05	12	–	1,600	62	55	GaAs, 5 V	1.2 W
[9]	ISSCC'05	12	–	1,700	64	50	0.35 um BiCMOS, 3 V	3 W
[10]	ISSCC'05	12	1/0.6	500	78	58	0.18 um CMOS, 1.8 V	216 mW
[11]	ISSCC'05	6	0.9/0.5	22,000	–	–	0.13 um BiCMOS, 3.3 V	1.2 W
[12]	ISSCC'04	14	1.8/0.8	1,400	–	60	0.18 um CMOS, 1.8 V	400 mW
[13]	ISSCC'04	10	0.1/0.1	250	74	60	0.18 um CMOS, 1.8 V	4 mW
[14]	ISSCC'04	14	0.65/0.55	200	85	44	0.18 um CMOS, 1.8 V	97 mW
[15]	ISSCC'03	16	1/0.25	400	95	73	0.25 um CMOS, 3.3 V	400 mW
[16]	ISSCC'03	14	0.43/0.34	100	82	62	0.13 um CMOS, 1.5 V	16.7 mW
[17]	ISSCC'01	12	0.3/0.25	500	75	35	0.35 um CMOS, 3 V	110 mW
[18]	ISSCC'00	14	0.5/0.5	100	82	72	0.35 um CMOS, 3.3 V	180 mW
[19]	ISSCC'99	14	0.3/0.2	150	84	50	0.5 um CMOS, 2.7 V	300 mW
[20]	ISSCC'99	14	0.5/0.5	60	85	75	0.8 um CMOS, 5 V	750 mW
[21]	ISSCC'98	10	0.2/0.1	250	71	57	0.5 um CMOS, 5 V	100 mW
[22]	ISSCC'98	12	0.6/0.3	300	70	40	0.5 um CMOS, 3.3 V	320 mW
[23]	VLSI'07	14	3.5/1	150	83	83	0.18 um CMOS, 1.8 V	127 mW
[24]	ESSCIRC'06	8	0.25/0.25	600	68	–	0.13 um CMOS, 1.2 V	2.4 mW

[25]	ESSCIRC'05	12	0.4/0.6	50	80	60	0.25 um CMOS, 3.3 V	270 mW
[26]	ESSCIRC'04	14	0.7/0.45	130	80	40	0.25 um CMOS, 3.3 V	103 mW
[27]	JSSC'06	5	–	32,000	31	30	300 GHz ft Bipolar	4.4 W
[28]	JSSC'06	12	0.38/0.44	180	72	62	0.25 um CMOS, 3.3 V	155 mW
[29]	JSSC'03	12	0.4/0.3	320	95	45	0.18 um CMOS, 1.8 V	60 mW
[30]	JSSC'03	14	0.3/0.3	300	72	68	0.25 um CMOS, 3.3 V	53 mW
[31]	JSSC'01	10	0.2/0.15	1,000	72	61	0.35 um CMOS, 3 V	110 mW
[32]	JSSC'98	12	0.6/0.3	300	70	40	0.5 um CMOS, 3.3 V	320 mW

[a] @550 MHz
[b] @186 MHz

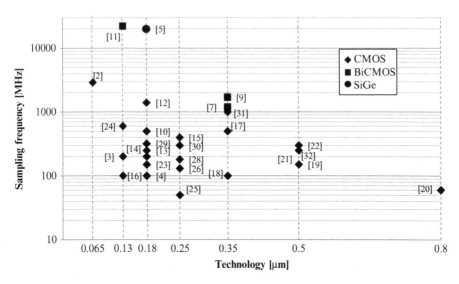

Fig. 2.13 State-of-the-art DACs: sampling frequency versus technology node

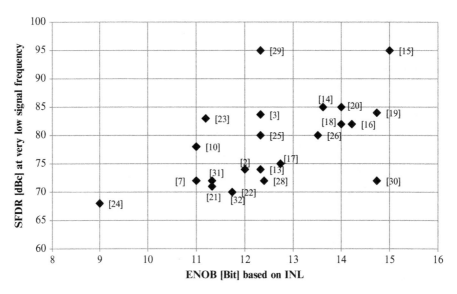

Fig. 2.14 State-of-the-art DACs: SFDR at very low signal frequencies (near DC) versus static ENOB

signal frequencies. For BiCMOS and III–V compounds technology, higher sampling and signal frequencies can be achieved, but the SFDR is still limited at 50 dBc around 1 GHz signal frequency.

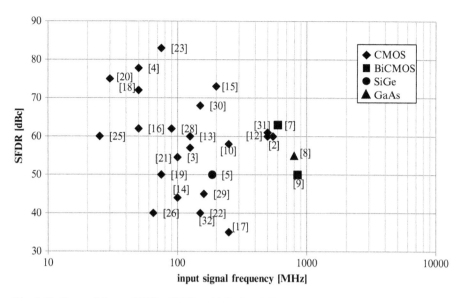

Fig. 2.15 State-of-the-art DACs: SFDR at high signal frequencies (near Nyquist) versus input signal frequency

Apparently, in order to achieve a good performance in a wide frequency range, it has to be analyzed how the DAC's static and dynamic performance are limited by various error sources. Accordingly, design techniques should be developed to overcome these design challenges. These issues are the main focus of this work.

2.6   Conclusions

In this chapter, the function and performance specifications of digital-to-analog converters (DACs) are briefly introduced. Different DAC architectures (binary, thermometer and segmented) and physical implementations (resistor, switched-cap and current-steering) are also discussed. The performance of state-of-the-art published DACs is summarized.

Due to its intrinsic high speed and driving ability, Nyquist current-steering DACs are most frequently used in high-speed, high-performance applications. Therefore, this work focuses on analysis and design techniques of current-steering DACs.

Chapter 3
Modeling and Analysis of Performance Limitations in CS-DACs

Dependent on where the errors are generated and how they affect the performance, errors in a current-steering DAC (CS-DAC) can be distinguished as non-mismatch errors (global errors) and mismatch errors (local errors). As mentioned in Sect. 2.4.3, regardless of whether the CS-DAC has a binary or thermometer or segmented architecture, it is composed of many current cells. If those current cells deviate from their ideal behavior differently, mismatch errors (such as amplitude and timing errors) are generated. If current cells perfectly match, i.e. no mismatch errors, non-mismatch errors, such as clock jitter, absolute duty-cycle error, finite output impedance and switching interferences, may still limit the DAC performance.

In this chapter, these mismatch and non-mismatch errors will be modeled and analyzed. The results of the analysis are confirmed by Matlab behaviorial-level simulations and are compared with other works. The achieved outcome gives a complete qualitative and quantitative overview of fundamental performance limitations for Return-to-Zero and Non-Return-to-Zero DACs, which will be the foundation to design a high-performance DAC.

In order to evaluate both amplitude and timing mismatch errors, i.e. to evaluate the dynamic-mismatch errors, two new parameters (the dynamic-DNL and dynamic-INL) are introduced to evaluate the dynamic matching between current cells. Compared to the traditional static-linearity parameters (the INL and DNL), the proposed dynamic-DNL and dynamic-INL describe the matching between current cells more completely and accurately. Based on this new concept, a novel smart design technique for the performance improvement will be developed in Chap. 5.

3.1 Static Mismatch Error

In this section, the static mismatch error of the current cells in a current-steering DAC, i.e. the amplitude error, will be discussed, including its effect on the DAC's static and dynamic performance.

Y. Tang et al., *Dynamic-Mismatch Mapping for Digitally-Assisted DACs*, Analog Circuits and Signal Processing 92, DOI 10.1007/978-1-4614-1250-2_3,
© Springer Science+Business Media New York 2013

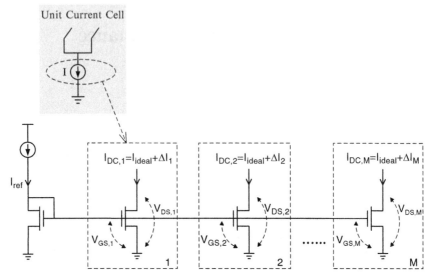

Current Reference   Current Sources in M Unit Current Cells

Fig. 3.1 Static mismatch error

3.1.1 Error Source: Amplitude Error

As described in Sect. 2.4.3, in an ideal thermometer-coded current-steering DAC, current sources in all unit current cells should provide the same static output current. These current sources are typically biased by a current mirror, as shown in Fig. 3.1. Assuming the transistors are operating in the saturation region, the static output current of a current source is given as:

$$I_{DC} = \frac{1}{2}\mu_n C_{ox} \frac{W}{L}(V_{GS} - V_{TH})^2(1 + \lambda V_{DS}) \qquad (3.1)$$

where μ_n, C_{ox} are the mobility of electrons and gate capacitance per unit area, respectively. λ is the channel-length modulation coefficient. V_{TH} is the threshold voltage, $V_{GS} - V_{TH}$ is the overdrive voltage and V_{DS} is the source-drain voltage. In practice, due to process and operating condition variations, current mismatches (ΔI_i) always exits between the mirrored currents ($I_{DC,i}$) of current sources. Variations in process parameters such as doping, gate-oxide thickness, lateral diffusion, oxide encroachment, and oxide charge density can drastically affect the electrical characteristics of a MOS transistor, which causes mismatches in μ_n, C_{ox} and V_{TH}. V_{TH} is also affected by the mechanical stress caused by the asymmetry of layout, such as shallow trench isolation (STI) stress. Operating conditions such as the overdrive voltage and V_{DS} can be affected by IR imbalance in the power supply network and by the environmental disturbance. In a word, all variations mentioned above contribute to the cell-dependent current mismatch (ΔI_i) between the DC-current of current cells.

In this work, the amplitude error (ΔA) of a current cell is defined as the ratio of the DC-current mismatch (ΔI) of this current cell over its ideal DC-current value. In general, since the overall gain error of a DAC does not have an negative impact on the DAC's performance, the ideal DC-current value can be considered as the mean DC-current value of all unit current cells. Equation (3.2) gives the amplitude error for each of M unit current cells shown in Fig. 3.1:

$$
\begin{aligned}
\Delta A_i &= \frac{\Delta I_i}{I_{ideal}} \\
&= \frac{(I_{DC,i} - I_{ideal})}{I_{ideal}} \\
&= \frac{I_{DC,i} - \frac{1}{N}\sum_{i=1}^{N} I_{DC,i}}{\frac{1}{N}\sum_{i=1}^{N} I_{DC,i}}, \ (i=1, 2, \ldots, M)
\end{aligned}
\tag{3.2}
$$

As discussed earlier in this section, the magnitude of ΔA is dependent on the process, circuit topology, transistor sizes, layout design, etc. Lots of mismatch models have been developed to investigate the transistor's parameters which can affect the current mismatch of current sources, such as electrical process parameters (threshold voltage V_t, current factor β, etc.) and the transistor size [32–35]. Though these references conclude with different models, the common point is that for a given technology, better intrinsic matching requires a larger transistor size. For example, as described in [32], the size of a transistor as a current source required to achieve ceratin matching accuracy is given as:

$$
\begin{aligned}
WL &= \frac{A_\beta^2 + \frac{4A_{VTH}^2}{(V_{GS}-V_{TH})^2}}{2(\frac{\sigma_I}{I})^2} \\
&= \frac{A_\beta^2 + \frac{4A_{VTH}^2}{(V_{GS}-V_{TH})^2}}{2\sigma_{amp}^2}
\end{aligned}
\tag{3.3}
$$

where W, L are the width and length of the transistor's gate. A_β and A_{Vt} are process-related proportionality parameters as defined in [32]. $V_{GS} - V_{TH}$ is the overdrive voltage. $\frac{\sigma_I}{I}$ is the relative standard deviation of current mismatch in current sources, which is equal to the standard deviation (σ_{amp}) of the amplitude error (ΔA) defined in (3.2). As seen from (3.3), the area of the current source has to be increased by a factor 4 for every extra bit of accuracy. The state-of-the-art accuracy achieved by non-calibrated DACs is 12–14 bit [2, 10, 12, 20, 29].

Since the amplitude error is cell dependent, i.e. input-data dependent, harmonic distortion will be generated such that both DAC's static and dynamic performance will be affected. The detailed analysis of the amplitude error's effect on the DAC performance are given in next two sections.

3.1.2 Effect on Static Performance

As introduced in Sect. 2.2.1, the differential non-linearity (DNL) and integral non-linearity (INL) are two parameters that show the static mismatch level of current cells and can be used to evaluate the static performance of DACs. Especially, the INL is most concerned since it directly affects the static linearity. Several models have been developed to investigate the relationship between σ_{amp} and the INL [32, 36, 37]. The most accurate reported model for the INL_{max} of a N-bit thermometer single-ended or differential DAC with Gaussian distributed amplitude errors, which is based on the min–max transfer curve explained in Sect. 2.2.1.1, is given in [37] as:

$$\mu_{INL_{max}} = 0.869\sigma_{amp}\sqrt{2^N - 1}\,\text{LSB}$$

$$\sigma_{INL_{max}} = 0.2603\sigma_{amp}\sqrt{2^N - 1}\,\text{LSB} \qquad (3.4)$$

where $\mu_{INL_{max}}$ and $\mu_{INL_{max}}$ are the mean and standard deviation of INL_{max}, respectively. As can be seen from (3.4), with fixed σ_{amp}, larger N means larger summed amplitude errors relative to a LSB. Therefore, the INL_{max} in LSB is approximately increased by $\sqrt{2}$ per extra bit in N. Note that if the INL_{max} is based on the best-fit transfer curve, it will be typically better than the result in (3.4).

3.1.3 Effect on Dynamic Performance

How amplitude errors affect the DAC dynamic performance will be discussed in this section. Amplitude errors are assumed to be Gaussian distributed. Single-Tone SFDR and THD are chosen to be analyzed. The analysis is based on statistical Monte-Carlo simulations in Matlab. The requirements on amplitude errors to achieve 3σ (99.7%) yield is also given.

3.1.3.1 Single-Tone SFDR/THD vs. Frequencies with Fixed Amplitude Error

Since amplitude errors belong to the class of static errors, the effect of amplitude errors has two general characteristics:

- The effect of amplitude errors is independent of frequencies, such as the sampling frequency (f_s) and the signal frequency (f_i). For a given amplitude error, due to

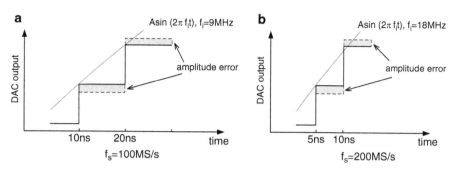

Fig. 3.2 Amplitude error with the same $\frac{f_i}{f_s}$ at different f_s

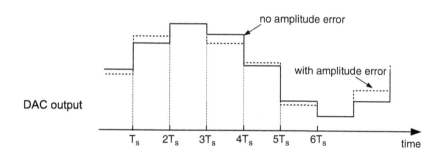

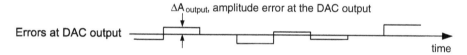

Fig. 3.3 Amplitude error at the DAC output

the fact that it is a static error, the effect of the amplitude error on the dynamic performance does not scale with the sampling frequency. As shown in Fig. 3.2, the amplitude error is the same for the same normalized input signal frequency $\frac{f_i}{f_s}$, such as 9 MHz@100 MS/s and 18 MHz@200 MS/s in this figure. Therefore, the dynamic performance, such as the THD and SFDR, will be the same for the same normalized input signal frequency. In addition, from the statistical perspective, if the power of the input sine-wave signal is constant, the power of the amplitude error should also be constant even with different input signal frequencies. As a result, the dynamic performance is also constant with the input signal frequency.

• With the same amplitude errors, the effects of amplitude errors on the performance of RZ and NRZ DACs are the same.

Figure 3.3 shows the output of an NRZ DAC with and without amplitude errors. T_s is the sampling period. Assuming the input signal is a sinusoid, for an N-bit

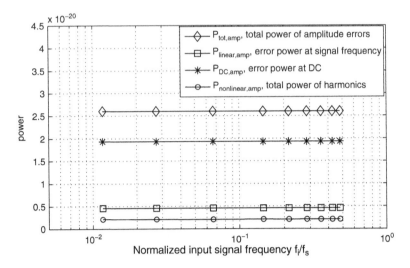

Fig. 3.4 Simulated power distribution of amplitude errors, mean value of 200 samples

thermometer-coded NRZ DAC, the rms value ($\Delta A_{output,rms}$) of the amplitude error at the DAC output can be approximated as:

$$\Delta A_{output,rms} \approx \frac{2A}{2^N - 1}\sigma_{amp}\sqrt{\frac{\frac{A}{\sqrt{2}}}{\frac{2A}{2^N-1}}} \tag{3.5}$$

where A is the peak amplitude value of the full-scale input sinusoid, σ_{amp} is the standard deviation of Gaussian distributed amplitude errors in current cells (relative to an ideal current cell, as defined in (3.2)).

Then, the total error power ($P_{tot,amp}$) generated by amplitude errors can be derived as:

$$P_{tot,amp} = \Delta A^2_{output,rms} = \frac{\sqrt{2}A^2}{2^N - 1}\sigma^2_{amp} \tag{3.6}$$

In fact, due to the actual error distribution and correlation to the input signal, part of the total error power ($P_{tot,amp}$) might be located at the input signal frequency and at DC. In other words, $P_{tot,amp}$ includes both the nonlinear error power $P_{nonlinear,amp}$, the linear error power $P_{linear,amp}$ (located at the signal frequency) and the error power $P_{DC,amp}$ (located at DC). To evaluate the DAC linearity, only the nonlinear error power ($P_{nonlinear,amp}$) needs to be considered. Figure 3.4 compares the simulated $P_{linear,amp}$, $P_{nonlinear,amp}$ and $P_{DC,amp}$ to $P_{tot,amp}$. As shown, $P_{nonlinear,amp}$ is only a small part of $P_{tot,amp}$. The ratio is about 9 dB.

Thus, assuming there is no sinc-attenuation at the DAC output, the THD in dBc, i.e. inverted signal-to-distortion ratio (SDR), can be calculated as:

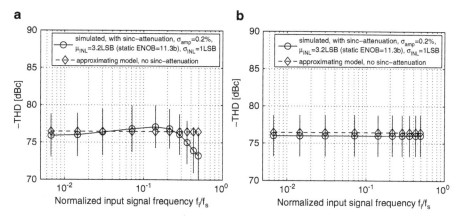

Fig. 3.5 THD vs. normalized input signal frequency at 500 MS/s. $\sigma_{amp} = 0.2\%$. Bars: one sigma spread (200 samples) (**a**) with sinc-attenuation (**b**) no sinc-attenuation

$$-THD_{amp,no-sinc} = SDR_{amp,no-sinc}$$

$$= 10 log_{10}(2^N - 1) - 10 log_{10}(\sigma_{amp}^2) + 4.5 \text{dBc} \quad (3.7)$$

As an example, Monte-Carlo statistical simulations with 200 samples are performed on a 14-bit 6T-8B segmented NRZ DAC. Since the thermometer part dominates the whole performance [38], the Gaussian distributed amplitude error is only assumed for the 6-bit thermometer part with a standard deviation $\sigma_{amp} = 0.2\%$, relative to an ideal thermometer current cell. The THD from Monte-Carlo simulations and the approximating model in (3.7), with input signal frequencies (f_i) normalized to a 500 MHz sampling frequency (f_s), are shown in Fig. 3.5. The simulated SFDR is shown in Fig. 3.6. The mean value of the simulated INL, based on the min–max transfer curve, is 3.2LSB for 14-bit accuracy. This equals to a static effective number of bits (ENOB) of 11.3. As seen in Figs. 3.5a and 3.6a, the simulated THD and SFDR are dependent on input signal frequencies. This is expected due to the sinc-attenuation given by the zero-order-hold shown in Fig. 2.2 for both signal and harmonic distortion:

- As the input signal frequency increases from very low frequencies, the harmonic distortion is moving to high frequencies faster than the signal itself (e.g. the second harmonic is moving 2x faster than the signal), resulting in a lager sinc-attenuation than for the signal. Therefore, the SFDR and THD are getting better.
- When the signal frequency increases further, the harmonic distortion moves out of the first Nyquist band and folds back, resulting in a less sinc-attenuation than for the signal. Therefore, the SFDR and THD are getting worse.
- If there is no sinc-attenuation at the DAC output, the SFDR and THD are statistically independent of frequencies, as shown in Fig. 3.5b. Then, as seen, the approximating model in (3.7) is well confirmed by the simulation results. The analysis result of the SFDR is also in line with [39].

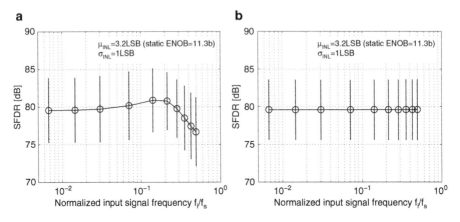

Fig. 3.6 SFDR vs. normalized input signal frequency at 500 MS/s. $\sigma_{amp} = 0.2\%$. Bars: one sigma spread (200 samples) (**a**) with sinc-attenuation (**b**) no sinc-attenuation

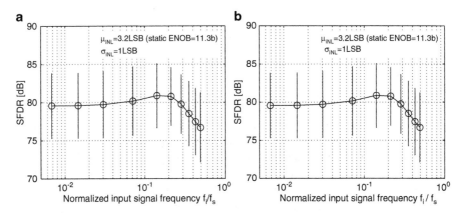

Fig. 3.7 SFDR vs. normalized input signal frequency at different sampling frequency, $\sigma_{amp} = 0.2\%$. Bars: one sigma spread (200 samples) (**a**) sampling frequency $f_s = 250$ MHz (**b**) sampling frequency $f_s = 1$ GHz

The effect of amplitude errors at different sampling frequencies is shown in Fig. 3.7. The sampling frequency is set to 250 MHz and 1 GHz, respectively. Compared to the previous results of 500 MS/s, with the same normalized input signal frequency, the SFDR is statistically independent of the sampling frequency. With different normalized input signal frequency, the SFDR is dependent on the sampling frequency due to the different sinc-attenuation. The same conclusion applies to the THD.

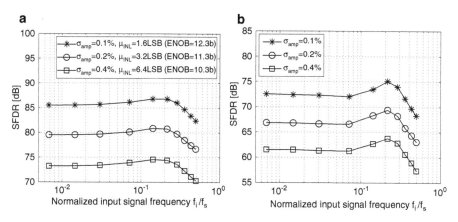

Fig. 3.8 SFDR vs. normalized input signal frequency with different σ_{amp}, 500 MS/s (**a**) SFDR mean value of 200 samples (**b**) SFDR 3σ (99.7%) yield curves of 200 samples

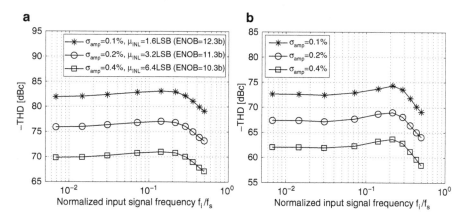

Fig. 3.9 THD vs. normalized input signal frequency with different σ_{amp}, 500 MS/s (**a**) THD mean value of 200 samples (**b**) THD 3σ (99.7%) yield curves of 200 samples

3.1.3.2 Single-Tone SFDR/THD vs. Amplitude Error with Fixed f_s

For given frequencies (f_s, f_i) and DAC architecture, larger amplitude errors cause larger harmonic distortion and more deterioration of the dynamic performance. Monte-Carlo statistical simulations (200 samples) of the relationship between the dynamic performance and the amplitude error were performed for this 14-bit 6T-8B segmented DAC. σ_{amp} was set to 0.1%, 0.2% and 0.4%, respectively. f_s is 500 MHz. Figures 3.8 and 3.9 show the simulated THD and SFDR with the mean value and 3σ (99.7%) yield curves. It can be seen that larger amplitude errors result in a worse dynamic performance: both SFDR and THD show a 20 dB/decade roll off with

Table 3.1 Summary of the effect of amplitude errors on the DAC performance

Variables	INL or DNL	SFDR or -THD
Amplitude error: σ_{amp}	-1 bit/octave	-20 dB/decade
Frequencies: f_s or f_i	–	Constant @ $\frac{f_i}{f_s} \leq 0.1$
		Depends on sinc-attenuation @ $\frac{f_i}{f_s} > 0.1$

σ_{amp} and the INL, or 6 dB per static effective bit. Regarding the yield, for example, in order to have at least 99.7% samples achieving > 60 dB SFDR in the whole Nyquist band, σ_{amp} should be smaller than 0.28% for this DAC example. Then, every extra 6 dB requirement on the SFDR or THD with the same yield requires σ_{amp} to be reduced by a factor of 2, independent on frequencies. According to (3.3), the area of a transistor as a current source has to be 4x larger in order to reduce σ_{amp} by half. Instead of just increasing the transistor size, more efficient design techniques to reduce the effect of amplitude errors will be discussed in Chaps. 4 and 5. The dependence of the DAC performance on amplitude errors and frequencies are summarized in Table 3.1 to give a clear overview.

3.2 Dynamic Mismatch Error

In the previous section, the static mismatch error (i.e. the amplitude error of current cells) and its effect on the performance has been discussed. However, in the applications with high signal and sampling frequencies, such as in communications and multimedia applications, the current cell is typically used as a switching current source rather than a static current source. Moreover, in those high-speed applications, the dynamic performance, such as SNDR, SFDR and IMD, are more important and concerned. With high signal and sampling frequencies, the dynamic performance is dominated by the dynamic switching behavior of current cells, instead of by their static behavior. Therefore, for high-speed high-performance DAC design, the mismatch between the dynamic switching behavior of current cells, called **dynamic mismatch**, has to be investigated.

3.2.1 Error Sources: Amplitude and Timing Errors

Figure 3.10a shows the dynamic switching behavior of current cells. The difference in the shape comes from mismatch in current sources, mismatch in switches, layout asymmetry, clock skew, unbalanced interconnection, etc. This mismatch is called **dynamic mismatch**. It is difficult to directly analyze the actual pulse output of current cells as shown in Fig. 3.10a. In order to build an efficient model,

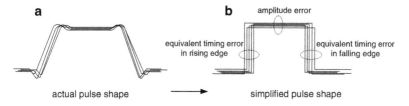

Fig. 3.10 Actual and simplified pulses

these pulses with complex shape have to be simplified. As shown in Fig. 3.10b, these pulses with complex shape can be translated into simple rectangular pulses with amplitude errors and equivalent timing errors in rising/falling edges. How to calculate the equivalent timing errors will be given in the next Sect. 3.2.1.1. Obviously, the dynamic mismatch between current cells is affected by both amplitude and timing errors. The effect of the static amplitude error on the DAC performance has already been discussed in Sect. 3.1. In this section, the timing error and its effect on the DAC performance will be covered.

3.2.1.1 Introduction to Timing Error

In order to simplify the pulses in Fig. 3.10a into rectangular pulses in Fig. 3.10b, both amplitude errors of rectangular waves and the equivalent timing errors in rising and falling edges of rectangular waves should be found. The amplitude errors can be easily found by measuring the static output current of the current cells. For the timing error, by using the method introduced in [40], an equivalent timing error can be found for a rising or falling edge.

As an example, Fig. 3.11 shows the way to define the equivalent timing error in the rising edge of a pulse. The equivalent timing error (t_r) in the rising edge is given by:

$$\Delta t_r = \frac{\Delta Q}{A} = \frac{\int_{t_1}^{t_2} \Delta I \, dt}{A} \tag{3.8}$$

where ΔI is the glitch error between the actual switching behavior of a current cell and the ideal switching behavior, t_2-t_1 is the duration of the glitch, A is the static amplitude of the output current of current cell, and ΔQ is the total charge of the error waveform. In fact, the shape of the glitch is not really of importance, since the duration of the glitch is very short compared to the sampling period of the DAC. The shape of the glitch only influences the very high frequency components. Only the total charge of the glitch error influences the frequency spectrum inside the band of interest. This simplification is accurate enough in most cases as verified in [40]. A similar definition can be used for the equivalent timing error in the falling edge (Δt_f). By this simplification, timing errors and amplitude errors are uncorrelated.

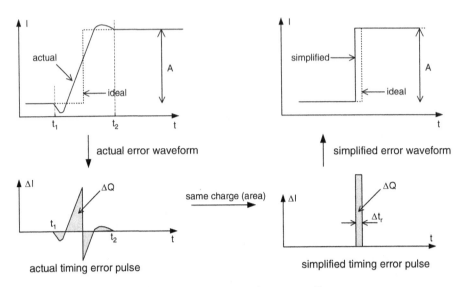

Fig. 3.11 Equivalent timing error in the rising edge of a current cell

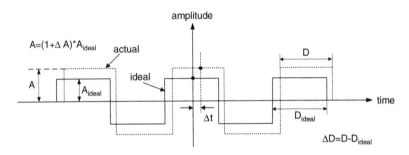

Fig. 3.12 Amplitude and timing (delay & duty-cycle) errors

In order to get more insight, the timing error can be decomposed into a delay error and a duty-cycle error, as shown in Fig. 3.12. The delay error (Δt) is defined as the timing difference in the middle point of the rectangular waves between the actual case and the ideal case, and the duty-cycle error (ΔD) is defined as the difference between the pulse width of the actual waveform and that of the ideal waveform. Note that the duty-cycle error (ΔD) referred in this work is an absolute error (in seconds), not a traditional relative error (a ratio in percentage). As shown in Fig. 3.13, the delay error (Δt) is typically caused by clock skew, mismatch in switches and latches, and interconnection and power supply imbalance between current cells, while the duty-cycle error (ΔD) is typically caused by the mismatch in the differential operation in individual current cells. Figure 3.14 shows an example of a duty-cycle error caused by the non-ideal differential switches (M1, M2) in a

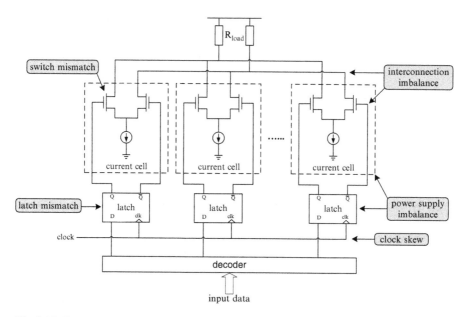

Fig. 3.13 Error sources of the delay error

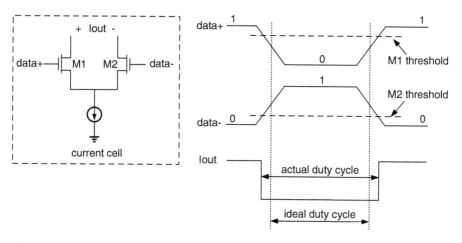

Fig. 3.14 A duty-cycle error caused by a threshold mismatch between differential switches

current cell. In Fig. 3.14, the threshold to turn M1 on/off is higher than the threshold
to turn M2 on/off. As can be seen, this threshold mismatch causes a duty-cycle error
at the output (Iout) of the current cell. In fact, any non-ideal differential operation,
such as mismatch between differential outputs of the latch, will cause duty-cycle
errors.

In summary, Δt is generated by the mismatch between the signal path of the current cells, while ΔD is generated by the mismatch inside the signal path of the current cells. Δt and ΔD can be derived from Δt_r and Δt_f as:

$$\Delta t = (\Delta t_r + \Delta t_f)/2 \tag{3.9a}$$

$$\Delta D = \Delta t_r - \Delta t_f \tag{3.9b}$$

It should be noted that the ideal switching pulse is the "averaged" switching pulse of all unit current cells, similar to the definition of the ideal DC-current value mentioned in Sect. 3.1.1. The common amplitude and delay of an ideal switching pulse only give a gain error and a latency of a DAC, but have no impact on the DAC's linearity, i.e. no impact on the DAC performance. However, an ideal pulse with non-zero common duty-cycle error still contributes to non-linearities and decreases the dynamic performance, such as SFDR, THD and IMD. This common duty-cycle error is a non-mismatch global error and should be minimized by design as much as possible. How this common duty-cycle error affects the DAC's dynamic performance will be discussed in Sect. 3.3.2. In this section, timing mismatch errors are assumed to be Gaussian distributed with zero mean, i.e. the common duty-cycle error is assumed zero.

Timing errors, including delay and duty-cycle errors, have no impact on the static performance (DNL and INL). However, since timing errors are local errors and current-cell dependent, timing errors will cause harmonic distortion and deteriorate the dynamic performance. Moreover, as will be explained in the next section, this effect will become more dominant at high sampling frequencies.

Since RZ DACs typically have a re-sampling stage at the output, RZ DACs do not suffer from timing mismatch errors, but still suffer from non-mismatch timing errors, such as clock jitter and the common duty-cycle error. Those errors will be analyzed for RZ DACs in Sect. 3.3.

The effect of linearly distributed timing mismatch error on the dynamic performance of NRZ DACs has been investigated in [41]. In the next section, the effect caused by Gaussian distributed timing mismatch errors on the dynamic performance of NRZ DACs will be analyzed.

3.2.1.2 Single-Tone SFDR/THD *vs.* Frequencies with Fixed Timing Error

Figure 3.15 shows the DAC outputs with and without timing errors, and the related timing error pulses. T_s is the sampling period. As seen, the magnitude of the timing error pulse is determined by the step size (y) of the DAC output transition. y depends on the difference between successive input digital codes (x):

$$y[nT_s] = x[nT_s] - x[(n-1)T_s] \tag{3.10}$$

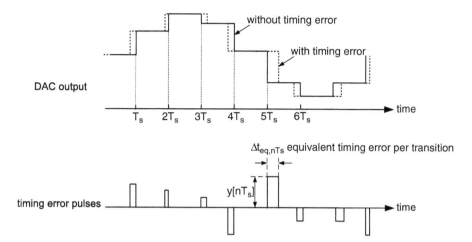

Fig. 3.15 Timing error pulses

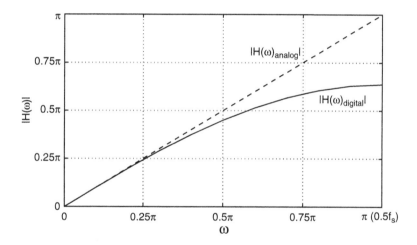

Fig. 3.16 Frequency response of first-difference analog and digital differentiators

Equation 3.10 behaves as a first-order discrete-time differentiator. This digital differentiator has a frequency magnitude response: $|H(\omega)_{digital}| = 2\sin(\frac{\omega}{2})$ [42], as shown in Fig. 3.16. The frequency axis in that figure covers the positive frequency range $0 \leq \omega \leq \pi$ samples/radian, corresponding to a cyclic frequency range of 0 to $0.5 f_s$, where f_s is the sample rate in Hz [42]. For comparison reasons, the frequency magnitude response of an ideal analog differentiator, $|H(\omega)_{analog}| = \omega$, i.e. the derivative of the input sine wave, is also shown in Fig. 3.16. As seen, $|H(\omega)_{digital}|$ matches $|H(\omega)_{analog}|$ accurately at very low frequencies. At high

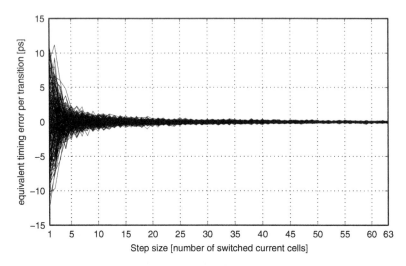

Fig. 3.17 Equivalent timing error per transition $\Delta t_{eq,nT_s}$, 200 samples

frequencies, $|H(\omega)_{digital}|$ is smaller than $|H(\omega)_{analog}|$. The relationship between $|H(\omega)_{digital}|$ and $|H(\omega)_{analog}|$ is:

$$|H(\omega)_{digital}| = |H(\omega)_{analog}| \times \frac{sin(\frac{f_i}{f_s}\pi)}{\frac{f_i}{f_s}\pi} \tag{3.11}$$

where f_i is the input signal frequency. Then, for a sine wave input signal $A sin(2\pi f_i t)$, the magnitude of timing error pulses can be derived as:

$$y[nT_s] = \frac{dAsin(2\pi f_i t)}{dt}\Big|_{t=nT_s} \times T_s \times \frac{sin(\frac{f_i}{f_s}\pi)}{\frac{f_i}{f_s}\pi} \tag{3.12}$$

The duration of the timing error pulse at the sampling moment nT_s is called equivalent timing error per transition ($\Delta t_{eq,nT_s}$). $\Delta t_{eq,nT_s}$ is calculated by averaging the timing errors of all current cells that need to be switched at the sampling moment nT_s:

$$\Delta t_{eq,nT_s} = \frac{\text{sum of the timing errors of the current cells that need to be switched}}{\text{number of the current cells that need to be switched}}$$

$$\tag{3.13}$$

Figure 3.17 shows $\Delta t_{eq,nT_s}$ in a 14-bit 6T-8B segmented DAC (200 samples), with the step size sweeping from one thermometer current cell to 63 thermometer current cells. The timing error is assumed as Gaussian distributed with the standard deviation $\sigma_{timing} = 5$ ps for the 6-bit thermometer part (the 8-bit binary part has no timing errors). As seen, due to averaging, $\Delta t_{eq,nT_s}$ is decreasing with the number of current cells that are switched together.

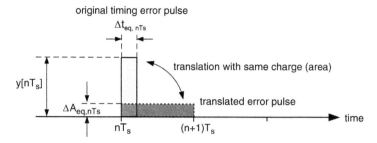

Fig. 3.18 Simplification of original timing error pulses

Typically, $\Delta t_{eq,nT_s}$ is within pico-second level and is far smaller than the sampling period (T_s). For the DAC performance in the first Nyquist band, only the low frequency components of timing error pulses are interesting. Therefore, to simplify the analysis, the error pulse generated by the timing error can be approximated to a new pulse which has a duration of one sampling period (T_s). As shown in Fig. 3.18, the equivalent amplitude error ($\Delta A_{eq,nT_s}$) of the translated error pulse is calculated by keeping the same charge as for the original timing error pulse. The reason to keep the same charge/area is to let the original and the translated pulses have the same low frequency components which are dominant in the performance. This simplification is safe in most applications for the DAC performance in the first Nyquist band.

Therefore, for a Non-Return-to-Zero (NRZ) DAC with a sinusoid input signal $A sin(2\pi f_i nT_s)$, the equivalent amplitude error ($\Delta A_{eq,nT_s}$) can be calculated as:

$$\Delta A_{eq,nT_s} = y[nT_s] \times \frac{\Delta t_{eq,nTs}}{T_s} = \frac{dAsin(2\pi f_i t)}{dt}\Big|_{t=nT_s} \times T_s \times \frac{sin(\frac{f_i}{f_s}\pi)}{\frac{f_i}{f_s}\pi} \times \frac{\Delta t_{eq,nTs}}{T_s}$$

$$= 2A\pi f_i cos(2\pi f_i nT_s) \times \frac{sin(\frac{f_i}{f_s}\pi)}{\frac{f_i}{f_s}\pi} \times \Delta t_{eq,nT_s} \qquad (3.14)$$

where f_i is the frequency of the input sinusoidal signal, A is its peak amplitude, f_s is the sampling frequency.

Then, for a N-bit thermometer NRZ DAC, the total error power ($P_{tot,timing}$) generated by timing errors can be derived as:

$$P_{tot,timing} = (2\pi f_i \frac{A}{\sqrt{2}} \frac{\sigma_{timing}}{\sqrt{\frac{\pi f_i (2^N-1)}{\sqrt{2} f_s}}} \sqrt{\frac{sin(\frac{f_i}{f_s}\pi)}{\frac{f_i}{f_s}\pi}})^2 \qquad (3.15)$$

where σ_{timing} is the standard deviation of Gaussian distributed timing errors in current cells.

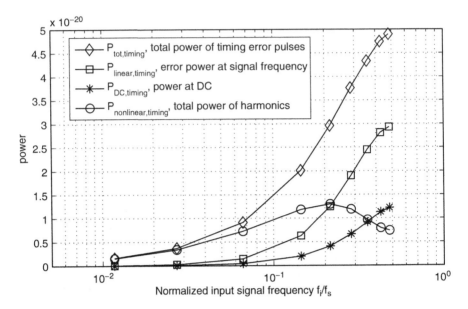

Fig. 3.19 Simulated power distribution of timing error pulses, mean value of 200 samples

Since $P_{tot,timing}$ is the total power generated by timing errors, it includes both the linear error power $P_{linear,timing}$ (located at the signal frequency) and the nonlinear error power $P_{nonlinear,timing}$. Also due to the actual error distribution, there may also be a part of error power located at DC ($P_{DC,timing}$). To evaluate the DAC linearity, only the nonlinear error power ($P_{nonlinear,timing}$) needs to be considered. Figure 3.19 compares the simulated $P_{linear,timing}$, $P_{nonlinear,timing}$ and $P_{DC,timing}$ to $P_{tot,timing}$. This Matlab simulation calculates the error power based on the original timing error pulses shown in Fig. 3.18, i.e. without simplification. As seen, when $f_i \ll f_s$, $P_{nonlinear,timing}$ almost equals to $P_{tot,timing}$, and $P_{linear,timing}$ is only a small part of $P_{tot,timing}$. This is because when $f_i \ll f_s$, due to the small step size, the equivalent timing error per transition ($\Delta t_{eq,nT_s}$) is highly random and largely spread as shown in Fig. 3.18, such that the timing error pulses are very little correlated to the signal. As $\frac{f_i}{f_s}$ increases, due to the increased probability of large step size, $\Delta t_{eq,nT_s}$ becomes smaller and the spread is less because of averaging, such that the timing error pulses are more correlated to the signal. This results in a large portion of the total error power being locating at the signal frequency. As seen in Fig. 3.19, as $\frac{f_i}{f_s}$ increases, the total error power ($P_{tot,timing}$) of timing errors pulses is more and more dominated by the error power at the signal frequency ($P_{linear,timing}$). $P_{linear,timing}$ is a linear error and should not be counted as harmonic distortion. The error power at DC ($P_{DC,timing}$) should not be counted in the distortion neither. By fitting, the simulation result in Fig. 3.19 shows an empirical correlation factor between $P_{nonlinear,timing}$ and $P_{tot,timing}$ as:

$$P_{nonlinear,timing} = (1 - 0.83sin(\frac{f_i}{f_s}\pi))P_{tot,timing} \qquad (3.16)$$

Then, the THD relative to the carrier, i.e. inverted signal-to-distortion ratio (SDR), can be calculated as:

$$THD_{timing} = \frac{1}{SDR_{timing}} = (\frac{P_{sig}}{P_{nonlinear,timing}})^{-1} = (\frac{\frac{A^2}{2}sinc^2(\frac{f_i}{f_s}\pi)}{P_{nonlinear,timing}})^{-1}$$

$$= \frac{4\sqrt{2}\pi\sigma_{timing}^2 f_i f_s}{(2^N - 1)sinc(\frac{f_i}{f_s}\pi)}(1 - 0.83sin(\frac{f_i}{f_s}\pi)) \qquad (3.17)$$

Note that since the duration of timing error pulses is very short compared to T_s, the sinc-attenuation for the distortion generated by timing error pulses is very small in the first Nyquist band. Therefore, as shown in (3.17), no sinc-attenuation for the distortion, only sinc-attenuation exists for the signal itself. This is different from the case of amplitude errors, where both the signal and the distortion generated by amplitude errors have sinc-attenuation.

The THD can also be expressed in dBc:

$$-THD_{timing,dBc} = SDR_{timing,dBc}$$

$$= 10log_{10}(2^N - 1) - 10log_{10}(\sigma_{timing}^2 f_i f_s \frac{1 - 0.83sin(\frac{f_i}{f_s}\pi)}{sinc(\frac{f_i}{f_s}\pi)}) - 12.5\,dBc$$

$$(3.18)$$

Monte-Carlo statistical simulations (200 samples) have been performed on a 14-bit 6T-8B segment DAC to verify (3.18). The Matlab simulation is based on original timing error pulses without simplification. The standard deviation σ_{timing} of timing errors is assumed to be 5 ps for the 6-bit thermometer part (the 8-bit binary part has no timing errors). Figure 3.20a compares the THD results of Matlab Monte-Carlo simulation and (3.18) at 500 MS/s. As shown, (3.18) is well confirmed by the simulation result. Unlike amplitude errors whose effect on the dynamic performance does not scale with frequencies, with given timing errors, the effect of timing errors on the dynamic performance does scale with frequencies. As shown in Fig. 3.20a, for $f_i \ll f_s$, -THD shows a 10 dB/decade roll off with f_i at a fixed f_s. The THD result for $f_i \ll f_s$ is similar to the result in [40] which is based on a power spectral density calculation. Compared to [40], this work also provided a simple equation for high $\frac{f_i}{f_s}$. As seen, when $\frac{f_i}{f_s}$ is larger than 0.2, the curve of -THD becomes flat with f_i, which is due to the correlation between the errors and the signal mentioned above. This conclusion also applies to the SFDR: as shown in Fig. 3.20b, the simulated SFDR has a -15 dB/decade roll-off with f_i until $\frac{f_i}{f_s} = 0.2$, then becomes flat. This phenomenon is also verified by the chip measurement results in Chap. 7.

The effect of timing errors at different sampling frequencies is easy to understand. When keeping the same normalized input signal frequency $\frac{f_i}{f_s}$, the proportion of

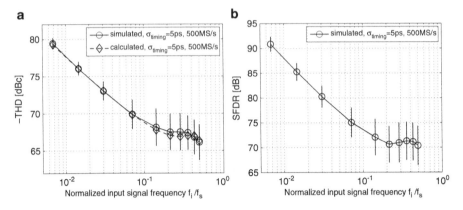

Fig. 3.20 THD, SFDR vs. normalized input signal frequency at 500 MS/s, $\sigma_{timing} = 5$ ps. Bars: one sigma spread (200 samples) (**a**) Calculated and simulated THD (**b**) Simulated SFDR

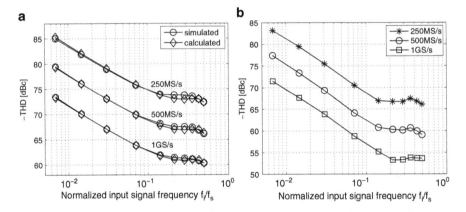

Fig. 3.21 THD vs. normalized input signal frequency at different sampling frequencies, $\sigma_{timing} = 5$ ps (200 samples) (**a**) Calculated and simulated THD mean value (**b**) Simulated THD 3σ (99.7%) yield curves

timing errors per sampling period doubles for doubled f_s. Therefore, the DAC's dynamic performance is expected to decrease 6 dB in that case. In other words, the dynamic performance should have a 20 dB/decade roll-off with f_s with the same normalized input frequency. This also means that as the signal and sampling frequencies increase, the effect of timing errors will dominate that of the amplitude errors. These expectations are confirmed by (3.18) and Matlab Monte-Carlo simulations at different sampling frequencies (250 MS/s, 500 MS/s and 1 GS/s) with $\sigma_{timing} = 5$ ps, as shown in Figs. 3.21 and 3.22. In those figures, both the mean value and 3σ (99.7%) yield curves of the SFDR and THD are shown. For example, in order to have at least 99.7% samples achieving >65 dB SFDR across the whole

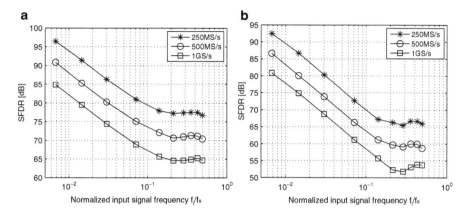

Fig. 3.22 SFDR vs. normalized input signal frequency at different sampling frequencies, $\sigma_{timing} = 5\,\mathrm{ps}$ (200 samples) (**a**) Simulated SFDR mean value (**b**) Simulated SFDR 3σ (99.7%) yield curves

Nyquist band at 250 MS/s, σ_{timing} should be smaller than 5 ps for this DAC example. Then, by keeping the ratio of $\frac{f_i}{f_s}$, every doubling of the sampling frequency will decrease the SFDR or THD by 6 dB.

3.2.1.3 Single-Tone SFDR/THD *vs.* Timing Error with Fixed f_s

It's easy to understand that for a given sampling frequency (f_s), signal frequency (f_i) and DAC architecture, larger timing errors cause larger harmonic distortion and deteriorate the dynamic performance. The expected result is a 20 dB/decade roll off with the timing error for both the THD and SFDR.

Results of (3.18) and Matlab Monte-Carlo simulations (200 samples) for this 14-bit 6T-8B DAC are shown in Figs. 3.23 and 3.24. Different Gaussian distributed timing errors are assumed for the 6-bit thermometer part with $\sigma_{timing} = 2.5\,\mathrm{ps}$, 5 ps and 10 ps, respectively. The sampling frequency f_s is 500 MHz. It can be seen that larger delay errors result in a worse dynamic performance: both SFDR and THD show a 20 dB/decade roll off with σ_{timing}. The results are also in line with the analysis in [40]. For example, as can be seen from the 3σ yield curves in Fig. 3.24, in order to have at least 99.7% samples achieving > 65 dB SFDR across the whole Nyquist band at 500 MS/s, σ_{timing} should be smaller than 2.5 ps. Then, every doubling of the timing error will decrease the SFDR or THD by 6 dB.

The dependence of the DAC performance on timing errors and frequencies are summarized in Table 3.2 to give a clear overview.

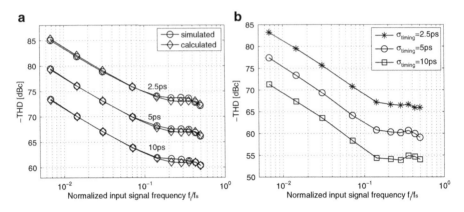

Fig. 3.23 THD vs. normalized input signal frequency with different σ_{timing} at 500 MS/s (200 samples) (**a**) Calculated and simulated THD mean value (**b**) Simulated THD 3σ (99.7%) yield curves

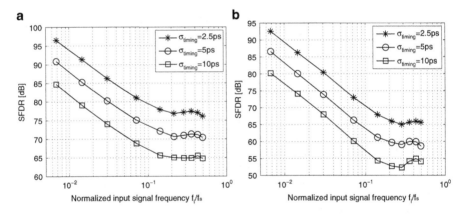

Fig. 3.24 SFDR vs. normalized input signal frequency with different σ_{timing} at 500 MS/s (200 samples) (**a**) Simulated SFDR mean value (**b**) Simulated SFDR 3σ (99.7%) yield curves

Table 3.2 Summary of the effect of timing error on the performance of NRZ DACs

Variables	-THD or SFDR
Timing error: σ_{timing}	-20 dB/decade
Frequencies: f_s or f_i	-20 dB/decade, when keeping the same $\frac{f_i}{f_s}$
	For different $\frac{f_i}{f_s}$, it depends on the actual situation as discussed

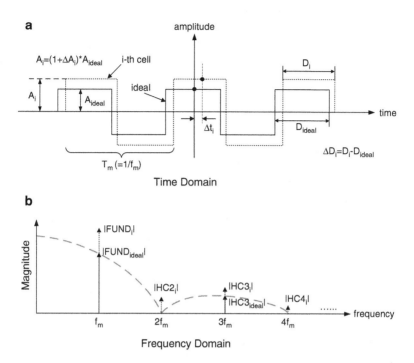

Fig. 3.25 Modulated rectangular output of current cells

3.2.2 Dynamic Mismatch in Frequency Domain

As we have seen, the dynamic mismatch is caused by both amplitude and timing errors. It can be dominated by the amplitude error or the timing error, depending on their values and the frequency. For low frequencies, the amplitude error is typically dominant in the dynamic mismatch, and the dynamic mismatch approaches the static mismatch. With increasing sampling and signal frequencies, the timing error becomes more and more visible. Consequently, for high-speed high-performance DAC design, the dynamic mismatch has to be investigated and minimized, rather than just minimizing the static mismatch.

The amplitude error can be measured by several methods, such as by means of current comparators and ADCs [14–16, 18, 25, 30]. However, measuring a timing error at pico-second level directly in the time domain is very difficult. In order to evaluate both amplitude and timing errors, measuring the dynamic mismatch in the frequency domain is an efficient way.

Instead of keeping the current cells at their static states to evaluate the static matching in the time domain, in this analysis of the dynamic matching, the outputs of the current cells are modulated as rectangular waves. Figure 3.25 shows the rectangular-wave output of an ideal unit current cell (without mismatch) and of the i-th actual unit current cell (with mismatch), in both time and frequency domain.

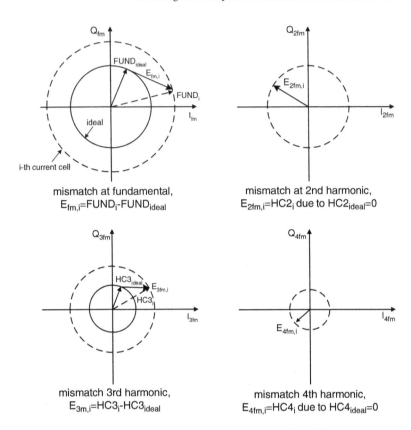

mismatch at fundamental,
$E_{fm,i}$=FUND$_i$−FUND$_{ideal}$

mismatch at 2nd harmonic,
$E_{2fm,i}$=HC2$_i$ due to HC2$_{ideal}$=0

mismatch 3rd harmonic,
$E_{3m,i}$=HC3$_i$−HC3$_{ideal}$

mismatch 4th harmonic,
$E_{4fm,i}$=HC4$_i$ due to HC4$_{ideal}$=0

Fig. 3.26 Dynamic mismatch in I-Q plane (for clarity, axis are not to scale)

A_i, D_i are the amplitude and pulse width of the output of the i-th unit current cell, respectively. ΔA_i, ΔD_i, Δt_i are respectively the amplitude error (relative to the ideal value), duty-cycle error and delay error of the i-th current cell, relative to the ideal current cell. f_m is the modulation frequency. As shown in Fig. 3.25b, because of the property of the rectangular wave, the magnitude of frequency components are shaped by a sinc function, where $FUND_i$, $HC2_i$, $HC3_i$, etc., are fundamental and harmonic components of the modulated output of the i-th cell. $FUND_{ideal}$, $HC3_{ideal}$, etc., are fundamental and odd harmonic components of the ideal current cell. For an ideal current cell, due to its duty-cycle error is zero, the even harmonics of the ideal cell are zero.

Then, dynamic-mismatch error E_{nfm} (n=1, 2, 3, ...) is defined as the mismatch error between an actual current cell and an ideal cell at the n-th order harmonic frequency, e.g. $E_{fm,i}$ and $E_{2fm,i}$ are the mismatch errors in the fundamental component $FUND_i$ and the second harmonic component $HC2_i$ of the i-th current cell, relative to an ideal cell, respectively. Since $E_{nfm,i}$ is a vector signal including both magnitude and phase information, it can be shown in an I-Q vector plane. Figure 3.26 shows the mismatch error $E_{nfm,i}$ (n=1, 2, 3, 4) of the fundamental up to the fourth harmonic components. How to measure the I-Q components will be given in Chaps. 5 and 6. Due to the property of the rectangular wave, for realistic mismatch values, the mismatch error ($E_{fm,i}$) at the fundamental frequency is dominated by the amplitude error (ΔA) and the delay error (Δt), and the mismatch error ($E_{2fm,i}$) at second harmonic is dominated by the amplitude error (ΔA) and the duty-cycle error (ΔD). Therefore, the mismatch errors at the fundamental and the second harmonic, i.e. $E_{fm,i}$ and $E_{2fm,i}$, already include all mismatch information and can represent the mismatch errors at all other harmonics. For instance, if $E_{fm,i}$ and $E_{2fm,i}$ are equal to zero, mismatch errors at all other higher-order harmonics are also equal to zero. Moreover, the mismatch errors at higher-order harmonic components are attenuated by the sinc function such that those mismatch errors are negligible compared to $E_{fm,i}$ and $E_{2fm,i}$. Therefore, it is enough to just use E_{fm} and E_{2fm} to evaluate the dynamic matching performance of current cells.

3.2.3 New Parameters to Evaluate Dynamic Matching: Dynamic-DNL and Dynamic-INL

As described in Sect. 2.2, the traditional DNL and INL are related to the differences between the ideal and the measured static transfer functions and are the parameters to evaluate the DAC's static performance. An example of a static transfer function is shown in Fig. 3.27. Since the static transfer function is only determined by the DC amplitude of the current cells, it is a one-dimensional curve with the digital input codes.

However, since both amplitude and timing errors together contribute to the dynamic non-linearity, just DNL and INL are not enough to evaluate the matching performance of current cells in current-steering DACs. Dynamic performance parameters, such as the SFDR, THD and IMD, only give a consequence of mismatch errors, but do not show the full relationship between the DAC's dynamic linearity and mismatch errors. In this section, two new parameters, called dynamic differential non-linearity (dynamic-DNL) and dynamic integral non-linearity (dynamic-INL), are introduced to evaluate the dynamic matching of current cells.

For the analysis of dynamic-INL and dynamic-DNL, a thermometer-coded DAC is chosen as an example. Dynamic-INL and dynamic-DNL can be derived based on the DAC's dynamic transfer function. As the digital input code is increased from zero to full scale, instead of measuring the static transfer function by sequentially turning on unit current cells as DC-current outputs, for the dynamic transfer function analysis, all unit current cells are sequentially turned-on and modulated as rectangular-wave outputs at frequency f_m. In other words: for the static transfer function, the DAC's output is a summed DC current of the current cells, while for the dynamic transfer function, the DAC's output is a sum of the modulated rectangular-wave output of the current cells. Figure 3.28 shows an ideal and an actual dynamic transfer function of the fundamental component of the modulated

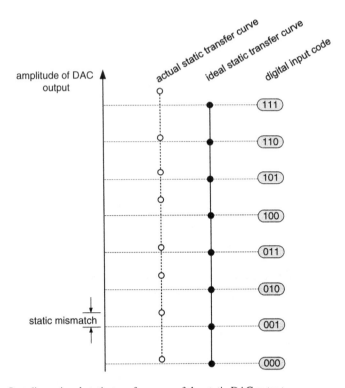

Fig. 3.27 One-dimensional static transfer curve of the static DAC output

DAC output in an I-Q plane. As shown, in the ideal case, with increasing input code, the fundamental component ($FUND_{ideal}$) of the DAC output linearly increases in magnitude, but keeps the same phase. However, in the actual case, due to amplitude and timing mismatch errors, the fundamental component ($FUND_{actual}$) has an non-linear increase in magnitude and also has a varying phase. Compared to the one-dimensional static transfer curve shown in Fig. 3.27, the dynamic transfer function includes both magnitude and phase information, and thus it is a two-dimensional curve with digital input codes. As seen, $E_{fm,code}$ (code=0~full scale digital input, i.e. 000, 001, ..., 111) is the integral non-linearity error of the fundamental component at a given code input. $E_{fm,code}$ is a sum of E_{fm} of the current cells that are turned-on at that code input, given as:

$$E_{fm,code} = FUND_{actual,code} - FUND_{ideal,code}$$

$$= \sum_{i=1}^{code} E_{fm,i}, \; \text{code=0~full scale digital input} \qquad (3.19)$$

The same definition of $E_{2fm,code}$ and $E_{3fm,code}$ can be used for the integral non-linearity errors of the second harmonic component (HC2) and the third harmonic component (HC3), and so on. For example, $E_{2fm,code}$ can be derived as:

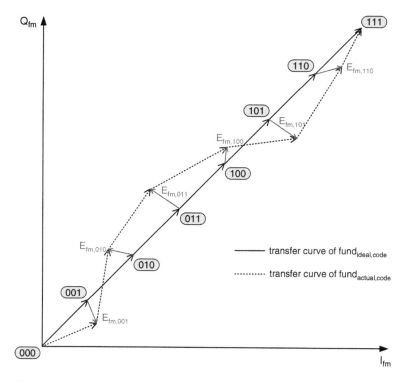

Fig. 3.28 Two-dimensional dynamic transfer curve of the fundamental component of the modulated DAC output

$$E_{2fm,code} = HC2_{actual,code} - HC2_{ideal,code}$$

$$= \sum_{i=1}^{code} E_{2fm,i}, \text{ code=0} \sim \text{full scale digital input} \qquad (3.20)$$

As mentioned in the previous section, E_{fm} and E_{2fm} already include all mismatch information and they are the most dominant two errors. Therefore, the dynamic integral non-linearity (dynamic-INL) in LSB can be defined as the RMS value of the summed power of the integral dynamic mismatch of the fundamental and second harmonic at every input code, i.e. summation of the power of $E_{fm,code}$ and $E_{2fm,code}$, over the power of the fundamental component of the modulated rectangular-wave output of one ideal LSB cell:

$$\text{dynamic-INL}_{code} = \sqrt{\frac{|E_{fm,code}|^2 + |E_{2fm,code}|^2}{|FUND_{ideal,1LSB}|^2}} \text{ LSB, code=0} \sim \text{full scale}$$

$$(3.21)$$

As seen, for optimization purposes, the dynamic-INL in (3.21) is converted back to a one-dimensional scalar with a unit of LSB, the same as the traditional static INL. Similar to the INL$_{max}$, the dynamic-INL$_{max}$ can be derived as:

$$\text{dynamic-INL}_{max} = max(\text{dynamic-INL}_{code}), \text{code=0}\sim\text{full scale} \qquad (3.22)$$

Similar to the DNL, the dynamic differential non-linearity (dynamic-DNL) is defined as the RMS value of the power of the differential dynamic mismatch at every input code, over the power of the fundamental component of the modulated rectangular-wave output of one ideal LSB cell:

$$\text{dynamic-DNL}_{code} = \sqrt{\frac{(|FUND_{code}|^2 + |HC2_{code}|^2) - (|FUND_{code-1}|^2 + |HC2_{code-1}|^2)}{|FUND_{ideal,1LSB}|^2}} - 1,$$

$$\text{code=1}\sim\text{full scale} \qquad (3.23)$$

Then, the dynamic-DNL$_{max}$ can be derived as:

$$\text{dynamic-DNL}_{max} = max(\text{dynamic-DNL}_{code}), \text{code=1}\sim\text{full scale} \qquad (3.24)$$

3.2.4 Comparison to Traditional Static DNL & INL

Compared to the traditional DNL and INL that are used to evaluate the static matching of current cells, the proposed dynamic-DNL and dynamic-INL include both amplitude and timing mismatch errors to evaluate the dynamic matching of switched current cells. Both the static and dynamic-switching behaviors of current cells are evaluated by these two new parameters. By reducing the dynamic-DNL and dynamic-INL, both amplitude and timing errors' effect on the DAC's static & dynamic performance will be reduced. While by only reducing the traditional DNL and INL, the DAC's static performance is improved and the DAC's dynamic performance is only partly improved since the timing errors still exist. This advantage of the dynamic-DNL and dynamic-INL over the traditional DNL and INL will be verified in Chap. 5. In conclusion, the dynamic-DNL and dynamic-INL describe the matching behavior of current cells more completely and efficiently than the traditional DNL and INL.

By definition, the dynamic-DNL and dynamic-INL are parameters not only related to amplitude and timing errors, but also modulation-frequency-dependent. Especially when the modulation frequency f_m is zero, the modulated rectangular-wave used to measure the dynamic-DNL and dynamic-INL becomes a 'DC' signal which is the same situation when measuring the traditional DNL and INL. As a result, when f_m is zero, the dynamic-DNL and dynamic-INL are equal to the traditional DNL and INL, respectively. Figure 3.29 shows an example of the

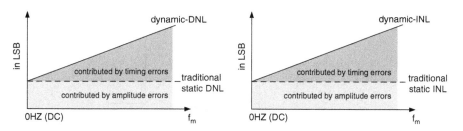

Fig. 3.29 dynamic-DNL, dynamic-INL *vs.* modulation frequency f_m

dynamic-DNL and dynamic-INL as f_m increases. As can be seen, at $f_m = 0$ Hz, the dynamic-DNL and dynamic-INL are equal to the traditional DNL and INL. This is obvious because at $f_m = 0$ Hz, the dynamic-DNL and dynamic-INL are determined only by the amplitude error. As f_m increases, the timing error comes in, and becomes more and more dominant in non-linearities.

In summary, if f_m changes, the weight between the amplitude error and the timing error in the dynamic mismatch changes, resulting in a frequency-dependent dynamic-DNL and dynamic-INL. Actually, this opens a door to find an optimized weight function between amplitude and timing errors for different applications to achieve the best performance. For example, more weight on the timing error for high sampling-rate applications, and more weight on the amplitude error for low frequency applications. How to improve the DAC's performance based on the selection of f_m and the optimization of the dynamic-DNL and dynamic-INL will be discussed in Chap. 5.

3.3   Non-mismatch Error

3.3.1   Sampling Jitter

Sampling jitter of a Digital-to-Analog Converter (DAC) is considered as the uncertainty in the updating time of the DAC output, as shown in Fig. 3.30. As can be seen, the jitter generates pulse-like errors in the DAC's output. The jitter can come from the device noise (thermal, 1/f), the power supply noise, interference cross-modulation, etc. It may include jitter from the clock source (phase noise) and jitter from the DAC itself. The jitter can be random or deterministic (such as sine or square-wave modulated), dependent on the mechanism of the jitter generation. If the sampling jitter is random and independent on the DAC's input data , the effect of the jitter behaves like white noise. In other words, the random sampling jitter does not generate distortion, but only increases the noise floor and thus decreases the signal-to-noise ratio (SNR). If the jitter is non-random and has a specific frequency component, inter-modulation distortion will be generated.

Fig. 3.30 Sampling jitter

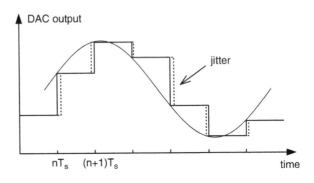

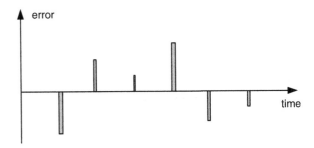

As mentioned in Sect. 2.1.1, ideal DACs do not generate quantization noise. In order to show the jitter effect from the DAC itself, the quantization noise at the input of the DAC is removed in the following analysis.

Since the jitter is also a kind of timing error, the translation of a timing error to an equivalent amplitude error introduced in Sect. 3.2.1.2 is used to convert the jitter error pulse into an equivalent amplitude error. Assuming the jitter is Gaussian randomly distributed, for a Return-to-Zero (RZ) DAC and a sine input signal, the equivalent amplitude error ($\Delta A_{RZ,nTs}$) of the translated jitter error pulse can be calculated as:

$$\Delta A_{RZ,nTs} = A\sin(\omega_i nT_s) \times \frac{\Delta t_{jitter,nTs}}{\frac{T_s}{2}}$$

$$= A\sin(2\pi f_i nT_s) \times 2\Delta t_{jitter,nTs} f_s$$

$$\sigma_{\Delta A_{RZ}} = \frac{A}{\sqrt{2}} \times 2\sigma_{jitter} f_s \qquad (3.25)$$

where f_i is the frequency of the input sinusoidal signal, A is its peak amplitude, f_s is the sampling frequency and σ_{jitter} is the variance of the jitter (rms value).

For a Non-Return-to-Zero (NRZ) DAC, the equivalent amplitude error ($\Delta A_{NRZ,nTs}$) of the translated jitter error pulse can be calculated as:

$$\Delta A_{NRZ,nTs} = \frac{d A sin(2\pi f_i t)}{dt}\Big|_{t=nT_s} \times T_s \times sinc(\frac{f_i \pi}{f_s}) \times \frac{\Delta t_{jitter,nTs}}{T_s}$$

$$= 2A\pi f_i cos(2\pi f_i nT_s) \times \Delta t_{jitter,nTs} \times sinc(\frac{f_i \pi}{f_s})$$

$$\sigma_{\Delta A_{NRZ}} = 2\frac{A}{\sqrt{2}}\pi f_i \times \sigma_{jitter} \times sinc(\frac{f_i \pi}{f_s}) \tag{3.26}$$

Then, the SNRs caused by the random jitter for a sinusoidal input signal in RZ and NRZ DACs can be calculated as:

$$SNR_{jitter,RZ} = \frac{P_{sig,RZ}}{P_{noise,RZ}} = \frac{\frac{A^2}{2}sinc^2(\frac{f_i \pi}{2f_s}) \times \frac{1}{2}}{(\frac{A}{\sqrt{2}}\sqrt{2}\sigma_{jitter}2f_s)^2 \times \frac{1}{2}} = \frac{sinc^2(\frac{f_i \pi}{2f_s})}{8\sigma_{jitter}^2 f_s^2} \tag{3.27a}$$

$$SNR_{jitter,NRZ} = \frac{P_{sig,NRZ}}{P_{noise,NRZ}} = \frac{\frac{A^2}{2}sinc^2(\frac{f_i \pi}{f_s})}{(2\pi f_i \frac{A}{\sqrt{2}}\sigma_{jitter}sinc(\frac{f_i \pi}{f_s}))^2} = \frac{1}{4\pi^2 \sigma_{jitter}^2 f_i^2}$$

$$\tag{3.27b}$$

For RZ DACs, the same Gaussian distributed jitter is assumed for both rising and falling edges. Due to the effect of return-to-zero, there are two jitter events in one sampling period for RZ DACs, resulting in a total jitter spread of $\sqrt{2}\sigma_{jitter}$, as shown in (3.27b).

The SNRs in decibels relative to the carrier for RZ and NRZ DACs can also be derived as:

$$SNR_{jitter,RZ} = -20log_{10}(\frac{\sigma_{jitter} f_s}{sinc(\frac{f_i \pi}{2f_s})}) - 9.03 \text{ dBc} \tag{3.28a}$$

$$SNR_{jitter,NRZ} = -20log_{10}(\sigma_{jitter} f_i) - 15.96 \text{ dBc} \tag{3.28b}$$

Equations (3.28a) and (3.28b) are verified by Matlab Monte-Carlo statistical simulations. The Matlab model calculates the SNR based on original jitter error pulses for RZ and NRZ DACs. The Matlab simulation results (1,000 runs) and Equations (3.28a), (3.28b) are all plotted in Fig. 3.31. The σ_{jitter} is 1 ps and 2 ps, respectively. The sampling frequency is 100 MHz and 200 MHz, respectively. Since the jitter is common for all current cells, the jitter effect is independent on the DAC's architecture. As shown, with Gaussian distributed random jitter, the theoretical model of (3.28a) and (3.28b) accurately match the Matlab simulation results. Figure 3.31 also shows that the translation error made during the translation from the jitter to the equivalent amplitude error does not affect the accuracy. The same analysis method can also be used for other types of jitter, such as sine-wave modulated jitter.

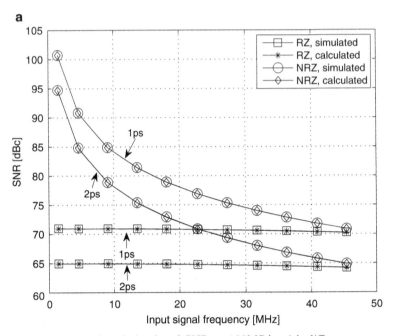

Calculated and simulated SNR at 100MS/s with different σ_{jitter}

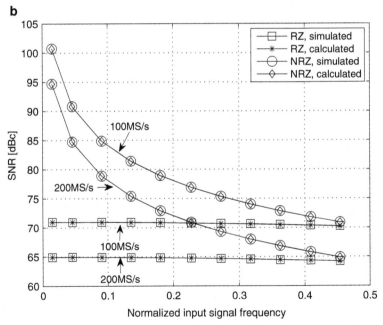

Calculated and simulated SNR with $\sigma_{jitter} = 1$ps for different f_s

Fig. 3.31 Jitter effect on the SNR of RZ and NRZ DACs

Table 3.3 Summary of jitter effects on DACs and ADCs

Variables	SNR of DAC		SNR of ADC
	RZ	NRZ	
σ_{jitter}	-20 dB/decade	-20 dB/decade	-20 dB/decade
f_s	-20 dB/decade	–	–
f_i	Almost constant, slightly affected by sinc-attenuation	-20 dB/decade	-20 dB/decade
Overall performance	☹	☺	

In summary, for RZ DACs:

- The SNR has a 20 dB/decade roll-off with σ_{jitter}.
- The SNR has a 20 dB/decade roll-off with the sampling frequency f_s. When f_s doubles, due to double switching, the noise power spectrum density (NSD) doubles. Together with the doubled noise bandwidth, it results totally in a 20 dB/decade roll off in the SNR.
- The SNR is slightly affected by the sinc-attenuation, otherwise it is independent of the signal frequency f_i. Because of the return-to-zero effect, the NSD is only dependent on the signal power, not the signal frequency. This is different from the jitter effect of the ADC, where the SNR has a 20 dB/decade roll-off with the signal frequency and is independent of the sampling frequency.

For NRZ DACs:

- The SNR has a 20 dB/decade roll-off with σ_{jitter}.
- The SNR is independent of the sampling frequency f_s. When f_s doubles, due to double switching and half transition step size, the NSD becomes half. Together with the doubled noise bandwidth, it results in a constant SNR.
- The SNR has a 20 dB/decade roll-off with the signal frequency f_i. When f_i doubles, the transition step size of DAC's output doubles, resulting in a quadruple NSD.
- The SNR is much better than the SNR of RZ DAC at low normalized frequencies (low f_i/f_s). This is because at low normalized frequencies, the amplitude of the transition step in a NRZ DAC is much smaller than that in a RZ DAC. Consequently, with the same jitter, the error in a NRZ DAC is also smaller than that in a RZ DAC at low normalized input frequencies. As a result, for those applications that have a critical noise requirement, a NRZ DAC will is a better choice.

Table 3.3 gives a straight-forward summary of jitter effects in DACs, with the jitter effect in ADCs as a reference.

Compared to the jitter analysis of DACs in literature [40, 43–45], the proposed analysis method is simple and straightforward, also with enough accuracy. References [43, 45] assume that the jitter analysis method of ADCs is also valid for NRZ DACs, but the proof is absent. In this work, based on a new and simple model, the analysis for both RZ and NRZ DACs are performed that are easy to understand. Previous research [40, 44] models the jitter effect based on a complicated PSD analysis of pulses, which gives the same result for NRZ DACs, but 0.9 dB difference for RZ DACs. This 0.9 dB difference for RZ DACs may come from the fact that the same jitter is assumed for both rising and falling edges in [40, 44] (i.e. the duration of a RZ pulse is fixed), while a different jitter is assumed between rising and falling edges in this proposed analysis (i.e. the duration of a RZ pulse is not fixed). In this work, a non-fixed RZ pulse duration is used by assuming that the jitter in rising and falling edges are uncorrelated.

3.3.2 Common Duty-Cycle Error

As mentioned in Sect. 3.2.1.1, common amplitude, delay and duty-cycle errors in current cells have different effects on the DAC's performance:

- A common amplitude error gives a gain error of the DAC, but it has no impact on the DAC's performance.
- A common delay error gives an output latency of the DAC. This also has no impact on the DAC's performance.
- A common duty-cycle error is a global non-mismatch error and will decrease the DAC's dynamic performance. As mentioned in Sect. 3.2.1.1, the duty-cycle error is caused by non-ideal differential operations in current cells. As a result, due to the common duty-cycle error, the DAC's dynamic performance will be suffer from a second harmonic distortion.

Figure 3.32 shows the common duty-cycle errors in thermometer-coded RZ and NRZ DACs. As can be seen, for a sinusoidal input signal $A sin(2\pi f_i t)$, the amplitudes of the error pulses caused by the common duty-cycle error for RZ and NRZ DACs follow the envelopes of $|A sin(2\pi f_i t)|$ and $|2\pi f_i T_s A cos(2\pi f_i t) sinc(\frac{f_i \pi}{f_s})|$, respectively. The Fourier series of $f(t) = |A sin(2\pi f_i t)|$ is given by:

$$f(t) = A\frac{4}{\pi}(\frac{1}{2} - \frac{1}{3}cos(2(2\pi f_i t)) - \frac{1}{15}cos(4(2\pi f_i t)) - \frac{1}{36}cos(6(2\pi f_i t)) -$$

$$\dots - \frac{1}{(2n)^2 - 1}cos(2n(2\pi f_i t))), -\pi \le 2\pi f_i t \le \pi \qquad (3.29)$$

Since the common duty-cycle error (D_{com}) is a common error for all current cells, similar to the jitter analysis in Sect. 3.3.1, after converting the common duty-cycle error into equivalent amplitude errors, the dominant second harmonic (HD2)

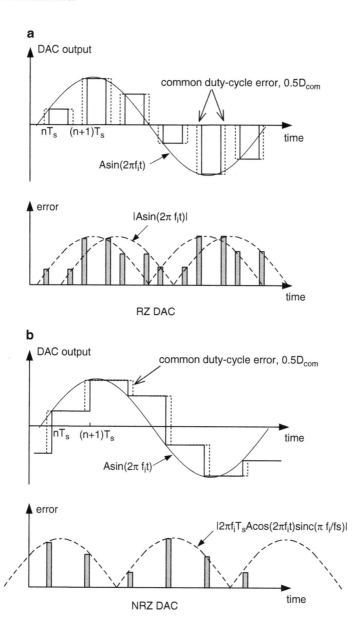

Fig. 3.32 Common duty-cycle error

relative to the carrier, caused by the common duty-cycle error for RZ DACs can be calculated according to (3.29):

$$HD2_{com-duty,RZ} = (\frac{P_{sig,RZ}}{P_{HD2,RZ}})^{-1} = (\frac{\frac{A^2}{2} \times sinc^2(\frac{f_i\pi}{2f_s}) \times \frac{1}{2}}{(\frac{A}{\sqrt{2}}\frac{4}{3\pi}\frac{|D_{com}|}{2} \times 2f_s)^2 \times (2 + 2|cos(\pi\frac{2f_i}{f_s})|) \times \frac{1}{2}})^{-1}$$

$$= \frac{16}{9\pi^2}D_{com}^2 f_s^2(2 + 2|cos(\pi\frac{2f_i}{f_s})|)\frac{1}{sinc^2(\frac{f_i\pi}{2f_s})} \qquad (3.30)$$

where D_{com} is the common duty-cycle error in seconds, and f_s, f_i are the sampling and input signal frequencies, respectively. According to the definition of the duty-cycle error, timing errors in rising and falling edges are equal to half of the common duty-cycle error ($\frac{D_{com}}{2}$). Moreover, as shown in Fig. 3.32a, for RZ DACs, due to the effect of return-to-zero, there are two identical error events in one sampling period. These two error events have the same envelope, but with a input-frequency-dependent phase shift. Thus, the total error power includes a correlation of $(2 + 2|cos(\pi\frac{2f_i}{f_s})|)$ for these two error events, as shown in (3.30). This also means that the HD2 will reach its minimal value when the input signal frequency is around $0.25 f_s$. This expectation will be proven by later Matlab simulations.

For NRZ DACs, the HD2 can be derived as:

$$HD2_{com-duty,NRZ} = (\frac{P_{sig,NRZ}}{P_{HD2,NRZ}})^{-1} = (\frac{\frac{A^2}{2}sinc^2(\frac{f_i\pi}{f_s})}{(2\pi f_i\frac{A}{\sqrt{2}}\frac{4}{3\pi}\frac{|D_{com}|}{2}sinc(\frac{f_i\pi}{f_s}))^2})^{-1}$$

$$= \frac{16}{9}D_{com}^2 f_i^2 \qquad (3.31)$$

With the common duty-cycle error, the SFDR is dominated by the HD2. Thus, the SFDR for RZ and NRZ DACs, in decibel, can be derived as:

$$SFDR_{com-duty,RZ} = -HD2_{com-duty,RZ}$$

$$= -10log_{10}(\frac{D_{com}^2 f_s^2(2 + 2|cos(\pi\frac{2f_i}{f_s})|)}{sinc^2(\frac{f_i\pi}{2f_s})}) + 7.44\,dB \quad (3.32a)$$

$$SFDR_{com-duty,NRZ} = -HD2_{com-duty,NRZ} = -10log_{10}(D_{com}^2 f_i^2) - 2.5\,dB$$

$$(3.32b)$$

Equations (3.32b) and (3.32b) have been verified by Matlab simulations, as shown in Figs. 3.33 and 3.34. The common duty-cycle error (D_{com}) is set to 1 ps, 2 ps and 4 ps, respectively. The sampling frequency is set to 200 MHz, 400 MHz and 800 MHz, respectively. As can be seen, the calculated theoretical model in Equations (3.32b) and (3.32b) match Matlab simulation results very well. The Matlab model calculates the SFDR/HD2 based on original error pulses caused by the common duty-cycle error, i.e. without converting the timing error into the

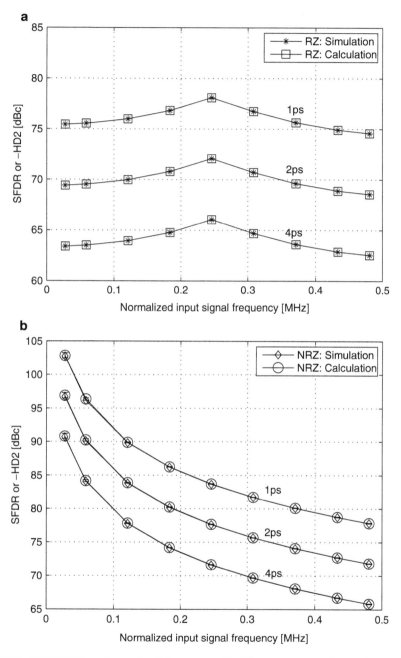

Fig. 3.33 SFDR/HD2 *vs.* input signal frequency with different common duty-cycle errors at 200 MS/s

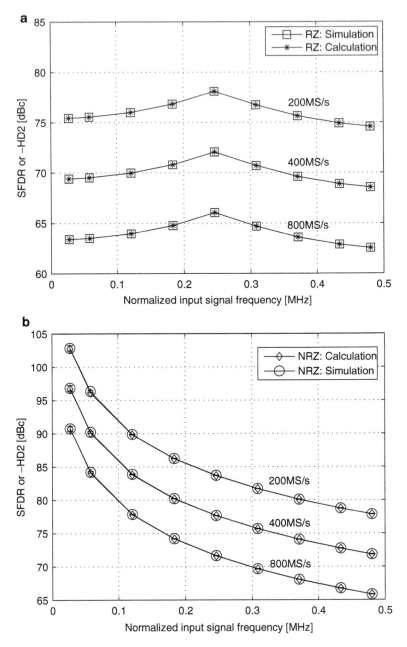

Fig. 3.34 SFDR/HD2 *vs.* normalized signal frequency with different sampling frequencies, $D_{com} = 1\,\text{ps}$

Table 3.4 Summary of the effect of the common duty-cycle error on the dynamic performance

Variables	SFDR or -HD2	
	RZ	NRZ
Common duty-cycle error: D_{com}	-20 dB/decade	-20 dB/decade
f_s	-20 dB/decade	–
f_i	Affected by correlation and sinc-attenuation as discussed	-20 dB/decade
Architecture: thermometer-bit N	–	–
Overall performance	☹	☺

equivalent amplitude error. Thermometer-coded RZ and NRZ DACs are chosen as examples. Since the common duty-cycle error is a common error existing in all current cells, the DAC architecture, i.e. the number of thermometer bits, has no influence on the DAC performance. The analysis results also apply to segmented DACs where the thermometer part is dominant in the whole DAC performance.

As can be seen from Fig. 3.33, the SFDR has a 20 dB/decade roll-off with D_{com} for both RZ and NRZ DACs. Moreover, as expected in (3.30), due to the correlation of error signals, the RZ DAC achieves the best SFDR when the input signal frequency (f_i) is around $0.25 f_s$. Compared to the SFDR when f_i is at very low or close to Nyquist rate, this best point of SFDR is 3 dB better. For NRZ DACs, the SFDR has a 20 dB/decade roll-off with f_i.

As shown in Fig. 3.34 where f_s is swept from 200 MHz to 800 MHz with normalized input frequencies, the SFDR has a 20 dB/decade roll-off with f_s for RZ DACs and is constant with f_s for NRZ DACs at the same absolute input frequency, respectively. The results are all in line with the derived theoretical equations (3.32b) and (3.32b). The effects of the common duty-cycle error for RZ and NRZ DACs are summarized in Table 3.4.

3.3.3 Finite Output Impedance

Another non-mismatch error that degrades the linearity of current-steering DACs is the signal-dependent output impedance [2,46]. As an example, a N-bit thermometer DAC with M ($= 2^N$-1) current cells is shown in Fig. 3.35. The number of current cells on each side of the differential output depends on the input signal x ($-1 \le x \le 1$), which equals $\frac{(1+x)M}{2}$ and $\frac{(1-x)M}{2}$ respectively. Since the output impedance (Z_o) of a current cell is not infinite, the total output impedance is signal-dependent and therefore the DAC's linearity decreases.

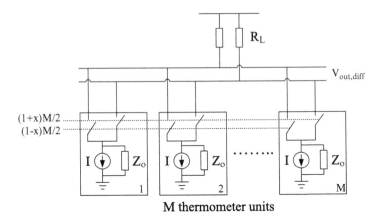

Fig. 3.35 Input-signal dependent output impedance

Assuming that a non-linear system is described as y(t)= $\alpha_1 x + \alpha_2 x^2 + \alpha_3 x^3$, the third-order harmonic distortion (HD3) and the third-order intermodulation distortion (IM3), which are most concerned in many communication systems, are given in [47]:

$$HD3 = 20log_{10}(\frac{\frac{\alpha_3}{4}}{\alpha_1 + \frac{\alpha_3}{4}}) \approx 20log_{10}(\frac{\alpha_3}{4\alpha_1}) \text{ dBc, for } x = \cos(\omega t)$$

$$IM3 = 20log_{10}(\frac{\frac{3\alpha_3}{4}(\frac{1}{2})^3}{(1/2)\alpha_1 + \frac{9\alpha_3}{4}(1/2)^2}) \approx 20log_{10}(\frac{3\alpha_3}{16\alpha_1}) \text{ dBc, for } x = \frac{1}{2}\cos(\omega_1 t)+\frac{1}{2}\cos(\omega_2 t)$$

$$(3.33)$$

$\alpha_1, \alpha_2, \alpha_3$ are assumed independent on the input signal. Compared to the HD3 analysis with a single input signal, half amplitude is assumed for both two input signals in the IM3 analysis. This is also a typical case in the practical IM3 measurement. According to (3.33), for the same non-linear system, the IM3 is 2.5 dB better than the HD3.

The analysis of the non-linear behavior due to the finite output impedance was introduced in [2]. The result of that analysis is given here as follows:

$$\begin{aligned}V_{out,diff} &= \frac{R_L I_o M(1 + x)Z_o}{R_L M(1 + x) + 2Z_o} - \frac{R_L I_o M(1 - x)Z_o}{R_L M(1 - x) + 2Z_o}\\ &= \frac{R_L M I_o}{2}\left(\frac{1 + x}{1 + \frac{2R_L M}{Z_o}(1 + x)} - \frac{1 - x}{1 + \frac{2R_L M}{Z_o}(1 - x)}\right)\end{aligned} \quad (3.34)$$

Substituting a = $R_L M/2Z_o$, $V_{out,diff}$ can be written as:

$$V_{out,diff} = \frac{R_L M I_o}{2}\left(\frac{1+x}{1+a(1+x)} + \frac{x-1}{1+a(1-x)}\right)$$

$$= \frac{R_L M I_o}{a^2}\left(\frac{x}{(1+\frac{1}{a})^2 - x^2}\right) \tag{3.35}$$

The Taylor expansion of (3.35) at x=0 is:

$$V_{out,diff} = \frac{R_L I_o M}{(a+1)^2}\left(x + \frac{x^3}{(1+\frac{1}{a})^2} + \ldots\right)$$

$$= \frac{R_L I_o M}{(\frac{R_L M}{2Z_o}+1)^2}\left(x + \frac{x^3}{(1+\frac{2Z_o}{R_L M})^2} + \ldots\right) \tag{3.36}$$

Based on (3.33), the HD3 can be derived as:

$$HD3 = 20log_{10}\left(\frac{\frac{\alpha_3}{4}}{\alpha_1 + \frac{3\alpha_3}{4}}\right) \text{ dBc} \approx 20log10\left(\frac{\alpha_3}{4}\right) \text{ dBc}$$

$$= 20log_{10}\left(\left(\frac{1}{2|1+\frac{2Z_o}{R_L M}|}\right)^2\right) \text{ dBc} \tag{3.37}$$

If Z_o is dominated by the output resistance (R_o) of the current cell at low frequencies, the HD3 can be approximated as:

$$HD3_{R_o} \approx 20log_{10}\left(\left(\frac{1}{2 + \frac{4R_o}{R_L(2^N-1)}}\right)^2\right) \text{ dBc} \tag{3.38}$$

As shown in (3.33), for the same non-linear system, the IM3 is 2.5 dB better than the HD3:

$$IM3_{R_o} = 40log_{10}\left(\frac{1}{2 + \frac{4R_o}{R_L(2^N-1)}}\right) - 2.5 \text{ dBc} \tag{3.39}$$

As seen, if Z_o is dominated by the resistive part R_o, the effect of the finite output impedance is independent of frequencies. In this case, the finite output impedance is a kind of static error. Figure 3.36 shows the minimal R_o required for -84 dBc IM3 performance with different thermometer bit N. R_L is assumed to be 50 ohm.

If Z_o is dominated by $1/j\omega C_o$ at high frequencies, where C_o is the effective output capacitance of the current cell defined in [2], the HD3 can be approximated as:

$$HD3_{C_o} \approx 20log_{10}\left(\left(\frac{1}{2|\frac{2}{j\omega R_L C_o M}|}\right)^2\right) \text{ dBc} = 40log_{10}\left(\frac{\pi f_i R_L C_o(2^N-1)}{2}\right) \text{ dBc}$$

$$\tag{3.40}$$

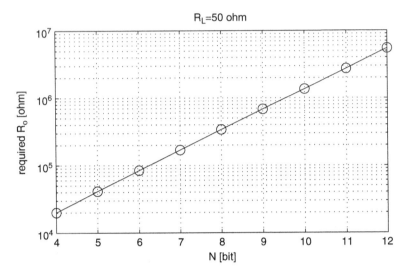

Fig. 3.36 Minimal R_o required for -84 dBc IM3 with different thermometer bit N

And the corresponding IM3 can be expressed as:

$$IM3_{C_o} = 40log_{10}(\frac{\pi f_i R_L C_o(2^N - 1)}{2}) - 2.5 \text{ dBc} \qquad (3.41)$$

In the case of Z_o dominated by the capacitive part at high frequencies, plots of the relationship between the HD3/IM3 and the input signal frequency (f_i) are shown in Fig. 3.37, where $R_L = 50$ ohm, $C_o = 6$fF and N = 6. As can be seen, the HD3 and IM3 have a 40 dB/decade roll-off with f_i. Therefore, for ultra high speed DACs, the finite output impedance will significantly decrease the dynamic performance and limit the SFDR/IMD. For example, depending on the process and circuit design, the output impedance starts to limit the SFDR and IMD at a signal frequency of 180 MHz in a 0.25 μm CMOS DAC [15] or 400 MHz in a 65 nm CMOS DAC [2].

3.3.4   Data-Dependent Switching Interference

As shown in Fig. 3.38, when switching a current cell, due to the parasitics and non-ideal switches, switching interference is generated:

- Depending on the crossover point of the differential data inputs of the switches (M1, M2), during the switching of the current from one side to another side, M1 and M2 may both be switched off or on for a short time. This both-OFF or both-ON states will cause a bounce in V_x at the common source node of M1 and M2. This voltage bounce will cause DAC non-linearities, e.g. the output current of the current source (M0) will be modulated by this voltage bounce.

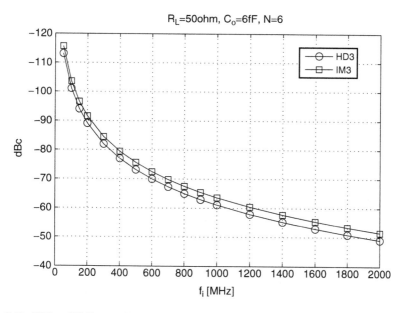

Fig. 3.37 HD3 and IM3 caused by finite output impedance versus f_i

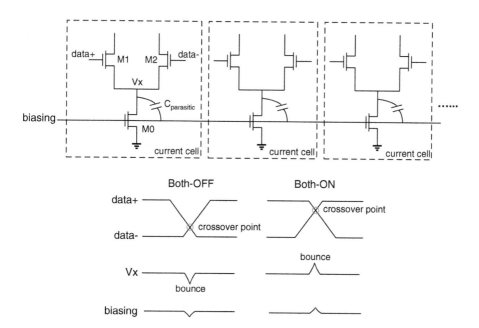

Fig. 3.38 Switching interference

- Due to parasitics coupling and common biasing, voltage bounces in one current cell might be coupled to other current cells. Then the output current of other current cells will also be modulated by this coupled voltage interference. Furthermore, the IR drops in the power supply network can also be affected by this voltage interference, resulting in a switching-dependent power supply for each current cell. Depending on the number of switching events at every sampling point, the strength of this disturbance is different and input-data dependent.

As mentioned above, switching interference is related to the switching events and are input-data dependent. Therefore, these data-dependent switching interference will generate harmonic distortion and decrease the dynamic performance, such as the SFDR and IMD. How severe this effect on the dynamic performance depends on the topology of the current cells, biasing scheme and shielding techniques. Several design techniques to reduce this effect will be discussed in Chap. 4.

3.4 Summary of Performance Limitations

Based on the error properties, the non-idealities in CS-DACs are categorized into three kinds of errors:

- Non-mismatch error: jitter, common duty-cycle error, finite output impedance, data-dependent switching interference;
- Static-mismatch error: amplitude error;
- Dynamic-mismatch error: (amplitude + timing) errors.

As discussed, both non-mismatch and mismatch errors limit the static and dynamic performance of DACs. Random sampling jitter contributes to white noise, which decreases the signal-to-noise ratio (SNR). In wireless communication applications where the requirement on noise performance is critical, a NRZ DAC is a better choice than a RZ DAC. Unlike the common amplitude or timing error, the common duty-cycle error is due to non-ideal differential operations and generates second harmonic distortion. Therefore, the common duty-cycle error has to be minimized by design as much as possible. Another non-mismatch error, finite output impedance, can become a dominant error source at high signal frequencies, such as above 180 MHz in a 0.25 μm CMOS DAC [15] or above 400 MHz in a 65 nm CMOS DAC [2], depending on the process and circuit design.

For low to intermediate signal frequencies (from DC to around 400 MHz), mismatch errors are still the typical dominant error sources in recently reported state-of-the-art CMOS DACs [2, 10, 12]. For near-DC signal frequencies, static errors, such as amplitude errors, are dominant. The dynamic performance shows a 20 dB/decade roll-off with the amplitude error. In addition, the effect of the amplitude error does not scale with frequencies: from a statistical point of view, the dynamic performance (such as the SFDR, THD and IM3) is constant with sampling and signal frequencies. However, the effect of the timing error does scale with frequencies: the dynamic performance shows a 20 dB/decade roll-off with the

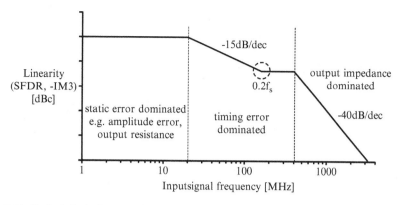

Fig. 3.39 Typical limitations on the DAC linearity by various error sources (fixed sampling frequency)

sampling frequency when keeping the same normalized input signal frequencies. As a result, as signal and sampling frequencies increase, the timing error will become more important and the effect of the timing error will exceed that of amplitude errors. Consequently, in order to achieve both good static and dynamic performance, both amplitude and timing errors have to be minimized. Figure 3.39 summaries how the DAC linearity, such as SFDR or IM3, is typically affected by various error sources for different input signal frequencies.

In this chapter, we simplified the mismatch errors of current cells into amplitude and timing errors. Since the timing error can be decomposed into delay and duty-cycle errors, it can be said that the dynamic mismatch is based on three-dimensional (**3-D**) mismatch errors (amplitude, delay and duty-cycle errors), in contrast to that the static mismatch is based on one-dimensional (**1-D**) mismatch error (amplitude error). Actually, in more practical situations, due to the dynamic mismatch being measured in the frequency domain, not only amplitude and timing errors, but all other kinds of mismatch errors are also included in the final measured results. That is to say, the dynamic mismatch and dynamic-DNL/dynamic-INL are based on all mismatch errors, or in other words, based on any-dimensional (**any-D**) mismatch errors. Table 3.5 gives a summarized comparison between the concepts of the traditional static mismatch and the proposed dynamic mismatch.

Moreover, as mentioned in Sect. 3.2.4, dynamic-DNL and dynamic-INL are frequency-dependent. This property of modulation-frequency dependency in the dynamic-DNL and dynamic-INL gives an advantage of choosing a suitable weight function between amplitude and timing errors, so that the DAC's performance can be optimized for different applications. For example, more weight on the timing error for high frequencies and vice versa. Based on the optimization of the dynamic matching, the non-linearities caused by both amplitude and timing errors can be decreased. As a result, both static and dynamic performance will be improved. Chapter 5 will introduce a new design technique to improve the performance, based on the optimization of the dynamic-INL.

Table 3.5 Comparison between static mismatch and dynamic mismatch

	Static mismatch (static matching)	Dynamic mismatch (dynamic matching)
Purpose	Evaluate the static behavior of current cells	Evaluate the dynamic behavior of current cells
Measured at	DC@time domain	f_m & $2f_m$ @frequency domain
Error covered	1-D: amplitude error	Any-D: all mismatch errors, or simplified 3-D: amplitude & timing (delay, duty-cycle) errors
Evaluation parameters	DNL, INL	Dynamic-DNL, dynamic-INL

3.5 Conclusions

In this chapter, the effect of various error sources on the DAC performance has been analyzed. Theoretical equations have been developed to describe the relation between these errors and the performance, and have been verified by Matlab simulations. The outcome provides guidelines to find bottlenecks of the performance during designing a high-performance CS-DAC.

Two novel design parameters, i.e. dynamic-DNL and dynamic-INL, have been introduced to evaluate the matching of dynamic switching behavior between current cells. The traditional static DNL and INL are only based on the amplitude error and are used to only evaluate the DAC's static performance. The proposed dynamic-DNL and dynamic-INL are based on both amplitude and timing errors. Therefore, the dynamic-DNL and dynamic-INL describe the matching between current cells more completely and accurately.

Moreover, the frequency-dependent characteristic of dynamic-DNL and dynamic-INL opens a door to choose a suitable weight function between amplitude and timing errors, so that the DAC's performance can be optimized for different applications. For example, more weight on the timing error for high frequencies and vice versa. Based on the optimization for the dynamic-DNL and dynamic-INL, the non-linearities caused by both amplitude and timing errors can be decreased. As a result, both DAC static and dynamic performance will be improved. A new design technique to improve the performance, based on the optimization of the dynamic-INL, will be introduced in Chap. 5.

Chapter 4
Design Techniques for High-Performance Intrinsic and Smart CS-DACs

In the previous chapter, the non-idealities in CS-DACs have been discussed and their impact on the DAC performance have been analyzed. In order to overcome these performance limitations, emerging design techniques for high-performance intrinsic and smart CS-DACs are introduced in this chapter. Firstly, the concept of smart DACs is introduced as intrinsic DACs with additional intelligence to acquire and utilize the actual chip information so that the performance/yield/reliability/flexibilty can be improved. Then, existing design techniques for intrinsic and smart DACs are discussed and summarized. Finally, a novel digital calibration technique for smart DACs, called dynamic-mismatch mapping (DMM), is initially introduced in this chapter and will be discussed further in the next chapters.

4.1 Introduction to Smart DACs

The smart concept implies on-chip intelligence to extract and use the actual chip information after manufacturing to improve the performance/yield/reliability/flexibi-lity beyond intrinsic limitations. As shown in Fig. 4.1, a smart DAC includes two main components:

- An intrinsic DAC.
- A feedback loop with on-chip smartness, including information sensors, process-ing circuits and actuators.

An intrinsic DAC is a basic DAC core designed based on a-priori knowledge. However, the actual chip information after manufacturing can not be accurately predicted by a-priori knowledge. On-chip intelligence in a smart DAC enables to have the actual chip information after chip production being fed back to the intrinsic DAC and utilize this information. The actual information of each chip is measured by information sensors. It might include error information, failure information, environment information, input signal information, user information,

Y. Tang et al., *Dynamic-Mismatch Mapping for Digitally-Assisted DACs*, Analog Circuits and Signal Processing 92, DOI 10.1007/978-1-4614-1250-2_4,

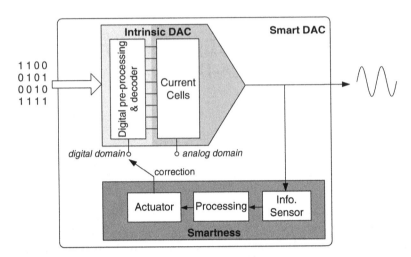

Fig. 4.1 Architecture of smart DACs

etc. After some necessary processing, actuators apply correction or optimization to the intrinsic DAC. This correction or optimization involves:

- Calibration of actual errors of current cells, including mismatch and non-mismatch errors, such that the DAC's performance and yield are improved.
- Replacement of current cells that fail due to some reasons (such as design errors, manufacturing failure, short life time, ESD breakdown, etc.) by redundant cells, so that the whole chip can still be functional. This will increase the chip's performance, yield, reliability and life time.

Compared to calibrating the errors of current cells, replacing a failure cell is rather easy. Therefore, this work focuses on the calibration and so on design techniques for high-performance DACs, i.e. design techniques that reduce the effect of non-idealities on the performance.

In smart DACs, the feedback loop to improve the DAC's performance is often called a calibration loop which includes error measurement, error processing and error correction. Since the DAC output is an analog signal, the error measurement has to be done in analog domain. Depending on where and how the correction from the actuator is applied, the calibration loop can be named as analog or digital calibration techniques :

- *Analog measurement, analog actuation* → analog calibration techniques
- *Analog measurement, digital actuation* → digital calibration techniques

For example, if the errors are corrected in the digital domain, i.e. the correction is done in digital pre-processing or decoding circuits, this calibration is called digital calibration technique; if the errors are corrected in the analog domain, i.e. the calibration is done in current cells, this calibration is called analog calibration technique. The comparison between analog and digital calibration techniques will be given in Sect. 4.3.2.1.

The fundamental reason why a smart DAC can potentially achieve a better performance than an intrinsic DAC is that a smart DAC measures, utilizes and calibrates the actual information of every chip so that every chip can achieve its own optimized performance. A smart DAC may retain this value for the rest of its useful life, or can be reconfigured for a different environment or application. In other words, a smart DAC provides an additional performance improvement or flexibility on an intrinsic DAC. How much performance improvement can be obtained depends on how large the errors in the intrinsic DAC are and how the on-chip smartness is implemented.

In the following sections, emerging design techniques for high-performance intrinsic DACs and smart DACs are introduced. After the analysis of these existing techniques, a novel calibration technique called dynamic-mismatch mapping (DMM), is introduced for smart DACs in the first place and is compared to other techniques. In Chap. 5, the proposed DMM technique will be discussed in detail and the implementation will be given in Chaps. 6 and 7.

4.2 Design Techniques for Intrinsic DACs

As discussed in Chap. 3, both non-mismatch and mismatch errors limit the performance of intrinsic DACs. In order to overcome these problems, many design techniques were proposed in recently published CS-DACs. Based on which error is focused on, these design techniques are divided into two groups: non-mismatch error related techniques and mismatch error related techniques.

4.2.1 Non-mismatch-Error Focused Techniques

As introduced in Chap. 3, non-mismatch errors mainly include sampling jitter, common duty-cycle error, finite output impedance and data-dependent switching interference. In this section, design techniques to minimize the effects of these non-mismatch errors in intrinsic DACs are discussed.

4.2.1.1 Sampling Jitter

As mentioned in Sect. 3.3.1, the sampling jitter in DACs has two sources:

- Jitter from clock sources, such as PLLs and oscillators.
- Jitter from the DAC itself.

Minimizing the jitter from clock sources requires low phase noise clock generators, such as low phase noise PLLs and oscillators. Design of a low phase noise clock generator is beyond the scope of this work which focuses on the DAC's design.

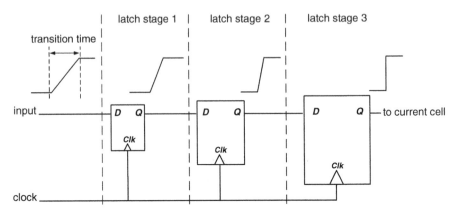

Fig. 4.2 Multi-stage clocked-latches to minimize the jitter generation

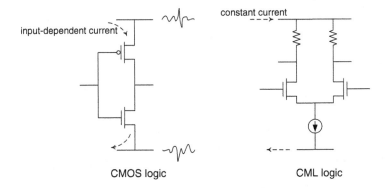

Fig. 4.3 CML logic *vs.* CMOS logic

Jitter from the DAC itself is mainly due to the noise in switches and power supplies at the switching moment. In order to minimize the jitter effect due to the noise in switches, the rule of thumb is to use the smallest transistor size in switches and large drive current, so that the transition time of switching can be made as short as possible, i.e. maximizing the $\frac{dV}{dt}$. Multiple clocked-latch stages are also often used to make the switching transitions as fast as possible in order to minimize the jitter generation, such as two latch stages in [10] and three latch stages in [2]. An example of three clocked latch stages is shown in Fig. 4.2. As can be seen, with increasing drive ability of latch stages, the transition time becomes shorter and shorter. Another benefit of multiple latch stages is that the timing mismatch error is also reduced due to clock re-sampling.

To minimize the jitter effect due to the noise from power supplies, the design of a clean, low-impedance power supply network is very important. If the requirement of the noise performance is highly concerned, it is recommended to use the current-mode logic (CML) in critical signal paths. As shown in Fig. 4.3, compared to CMOS

Fig. 4.4 Simple cascoding

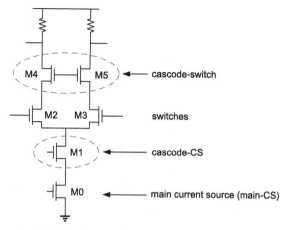

current cell with simple cascoding

logic, CML logic has a constant, input signal independent current consumption, resulting in a lower noise injection and less cross-talk to power supplies and the substrate [10].

4.2.1.2 Common Duty-Cycle Error

As mentioned in Sect. 3.3.2, the common duty-cycle error is due to non-ideal differential operations in DACs. It generates second-order distortion and limits the wide-band performance of the DAC. However, in some narrow-band applications where the third-order intermodulation (IM3) is most important, such as multi-carrier OFDM communication systems, the second-order intermodulation components (IM2) are out of the signal band. Therefore, in those applications, the effect of the common duty-cycle error is not of too much concern.

Current design techniques to minimize this common duty-cycle error in intrinsic DACs still rely on the intrinsic matching of the transistors and the matching on the layout. Measuring the common duty-cycle error is also difficult since it is at picosecond or even sub-picosecond level. Up to now, in any published DAC, there is no such a technique or concept proposed yet to measure or calibrate this common duty-cycle error.

4.2.1.3 Finite Output Impedance

Regarding the minimization of the impact of finite output impedance, simple cascoding [10, 15, 17, 20] and always-on cascoding [2, 48] were developed.

As shown in Fig. 4.4, simple cascoding is widely used to increase the output impedance of current cells, especially at low signal frequencies where the output resistance is dominant. As is well known, the output resistance is improved by $g_m r_o$ per cascode stage.

Fig. 4.5 Half-cell circuit at M2 on/off state

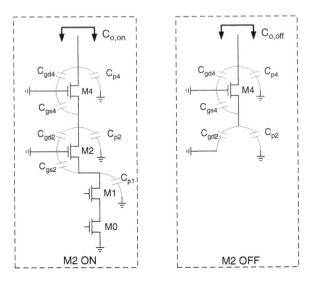

However, at high signal frequencies, simple cascoding can not help too much. This is because at high frequencies, the output capacitance (C_o) is dominant in the output impedance (Z_o) of the current cells. A half-cell circuit of Fig. 4.4 is shown in Fig. 4.5. Depending on the ON or OFF state of the switch M2, the output impedance of the current cell at high frequencies can be derived as [2]:

$$Z_{o,on} \approx \frac{1}{j\omega C_{o,on}} = \frac{1}{j\omega\left[(C_{gd4} + C_{p4} + \left(\frac{C_{gs4}+C_{gd2}+C_{p2}}{g_m r_o}\right)_{on} + \left(\frac{C_{gs2}+C_{gd1}+C_{p1}}{(g_m r_o)^2}\right)_{on} + \cdots\right]}$$

$$Z_{o,off} \approx \frac{1}{j\omega C_{o,off}} = \frac{1}{j\omega\left[(C_{gd4} + C_{p4} + \left(\frac{C_{gs4}+c_{gd2}+C_{p2}}{g_m r_o}\right)_{off}\right]} \tag{4.1}$$

where C_{gd} is the gate-drain capacitance, C_{gs} is the gate-source capacitance, and C_p is the parasitic capacitance in the source and drain. For simplification, these capacitors are assumed voltage-independent and all transistors are assumed to have the same gain $g_m r_o$.

As can be seen from (4.1), since the ON or OFF state is decided by the input signal, the difference ($C_{o,diff}$) between $C_{o,on}$ and $C_{o,off}$ results in a signal-dependent output impedance:

$$C_{o,diff} = C_{o,on} - C_{o,off}$$
$$= \left(\frac{C_{gs4} + C_{gd2} + C_{p2}}{g_m r_o}\right)_{on} - \left(\frac{C_{gs4} + C_{gd2} + C_{p2}}{g_m r_o}\right)_{off}$$
$$+ \left(\frac{C_{gs2} + C_{gd1} + C_{p1}}{(g_m r_o)^2}\right)_{on} + \cdots \tag{4.2}$$

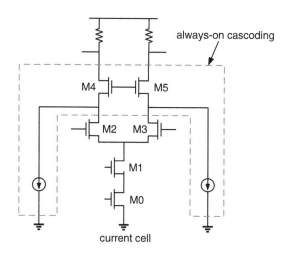

Fig. 4.6 Always-on cascoding

where $(\frac{C_{gs4}+C_{gd2}+C_{p2}}{g_m r_o})_{on} - (\frac{C_{gs4}+C_{gd2}+C_{p2}}{g_m r_o})_{off}$ is typically the dominant part. For example, if M4 is also totally turned off during the OFF state of M2, $(\frac{C_{gs4}+C_{gd2}+C_{p2}}{g_m r_o})_{off}$ will be zero.

Adding another cascoding stage on top of M4 and M5 does not help because the signal-dependent output impedance is always dominated by the output capacitance of the most top cascode stage. In order to improve the output impedance at high frequencies, a special cascoding called always-on cascoding was developed in [2, 48], as shown in Fig. 4.6. By always conducting a certain current in M4 and M5, i.e. by making $(\frac{C_{gs4}+C_{gd2}+C_{p2}}{g_m r_o})_{on}$ more equal to $(\frac{C_{gs4}+C_{gd2}+C_{p2}}{g_m r_o})_{off}$, $C_{o,diff}$ can be significantly reduced. Then, the output impedance becomes more independent of the signal. Therefore, compared to simple cascoding, always-on cascoding can reduce the distortion due to the signal-dependent output impedance also at high signal frequencies.

4.2.1.4   Data-Dependent Switching Interference

As mentioned in Sect. 3.3.4, if the switching interference is data-dependent, it will limit the DAC's dynamic performance. For intrinsic DACs, in order to minimize the effect of the switching interference generated during cell switching, a switching scheme called constant switching is proposed to make the switching interference data-independent [12, 49], as shown in Fig. 4.7. The constant switching scheme guarantees that there is always a switching action at every sampling clock in each current cell, i.e. the current (I_{cs}) interchanges its path between path A and path B at every sampling clock. However, the output current (I_o) of the current cell is still controlled by the input data (D1 to D4) of the switches (M1 to M4), i.e. the sign of I_o will change or stay the same. By doing this, the switching interference is not input-data dependent anymore, but becomes clock-frequency dependent. As shown in Fig. 4.7, the distortion caused by the switching interference after constant switching will appear at the clock frequency, which is unimportant for most applications.

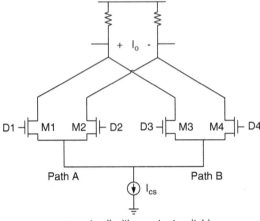

current cell with constant switching

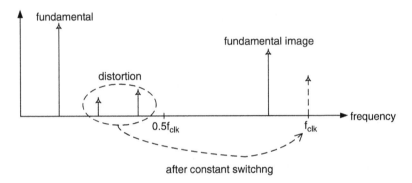

after constant switchng

Fig. 4.7 Constant switching scheme

Besides constant switching to shift the switching interference to the clock frequency, another technique is also developed in [12] to really reduce the switching interference by controlling the crossover point of input data signal to the switches. Since there is a feedback loop in this technique, it will be discussed in Sect. 4.3.1.1, as a design technique for smart DACs. Other basic design techniques to reduce switching interferences for intrinsic DACs include local biasing, low impedance power supply and good layout shielding between current cells.

4.2.1.5   Design Techniques for Multiple Non-mismatch Errors

Though the DAC's non-linear behavior is caused by many error sources, if the DAC is considered as a black box with certain non-linear behavior, techniques can be developed to minimize the whole non-linear behavior of the DAC. Harmonic suppression is one of these techniques, as proposed in [50, 51].

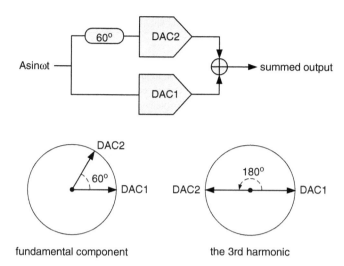

Fig. 4.8 Harmonic suppression

The basic idea of harmonic suppression with a sine-wave input example is shown in Fig. 4.8. Assuming DAC1 and DAC2 have the same non-linear behavior, after shifting the input signal of one DAC by 60°, the third harmonic is shifted by 180° and will be canceled by that of the other DAC. However, in this harmonic suppression technique, the same non-linear behavior is assumed for both DACs. That is to say, harmonic suppression technique is particularly useful for global/systematic non-mismatch errors [50,51]. However, for random mismatch errors, two DACs do not have the same non-linear behavior. Therefore, harmonic suppression technique with a fixed 60° phase shift will not improve so much in that case, unless the non-linear behavior of all sub-DACs is measured and certain algorithms are developed to cancel the errors.

4.2.2   Mismatch-Error Focused Techniques

To overcome the performance limitations caused by mismatch errors mentioned in Chap. 3, minimizing mismatch errors in the design of intrinsic DACs is typically based on following basic design rules:

- Properly size and bias the transistors in current sources to minimize the amplitude error.
- Synchronize the data with the same clock and make the transition time of the switching signal as short as possible to minimize the timing error.
- Make the layout as symmetrical as possible to minimize both amplitude and timing errors.

Table 4.1 Summary of advanced design techniques for intrinsic DACs

Reference	Targeted mismatch error		Techniques
	Amplitude	Timing	
[30, 32]	O		Common-centroid layout
[31]	O		Double common-centroid layout
[17]	O		Triple common-centroid layout
[19, 52]	O		Q^2 Random-Walk based switching sequence
[10, 40]	O		Stochastic based switching sequence
[41]		O	Delay-difference cancelation based switching sequence
[4, 23, 53]	O	O	Dynamic element matching (DEM)

The size of the transistors in current sources required for certain accuracy is already discussed in Sect. 3.1.1. The required size depends on the quality of the process and the required accuracy. Besides sizing, the biasing for current sources is also very important. Local biasing helps to reduce the biasing mismatch, as introduced in [40]. In order to minimize the timing error, strong drivers, multiple latch stages and the symmetrical layout are often used in the clock and switching signal path [2, 10], as shown in Fig. 4.2.

Besides the basic rules, various advanced techniques were also proposed to improve the performance of intrinsic DACs [4, 10, 17, 19, 23, 30–32, 40, 41, 52, 53], as summarized in Table 4.1.

The layout techniques introduced in previous work [10, 17, 19, 30–32, 40, 52] address amplitude errors caused by systematic process gradient errors. Those works decompose each unit current source into several sub-elements and find an optimized way to re-group these sub-elements into new unit current sources, so that the amplitude mismatch error is reduced. The number of sub-elements is typically an integer multiple of 2, such as 4 in [31, 32, 40] and even 32 in [30]. The more sub-elements, the better error averaging, but more complex layout and larger chip area. Since the DAC's static linearity is determined by the accumulated amplitude errors, the switching sequence of current cells is also optimized to improve the INL caused by the systematic amplitude error [10, 19, 40, 52] or to cancel the systematic delay errors [41]. However, those techniques are all based on the estimated systematic errors (i.e. process spatial gradient errors) and can not reduce random amplitude or timing errors. Since mismatch errors are random in a good design, most design techniques optimize those random errors based on-chip error measurement. Those design techniques to reduce random amplitude and timing errors belong to the design techniques of Smart DACs and will be discussed in the next section.

Beside layout techniques, another famous circuit design technique, called dynamic element matching (DEM) or dynamic averaging, is proposed to optimize the effects of both amplitude and timing errors [4, 23, 53]. DEM is widely used as a digital performance-enhancement technique in many electronic designs, such as in ADCs, DACs and sensors. The basic idea of DEM for DACs is randomizing the way of using circuit elements, e.g. randomizing the switching sequence of current cells,

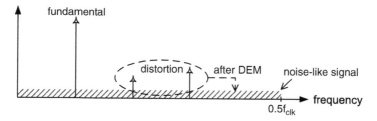

Fig. 4.9 Spectrum spreading by DEM

so that no signal-dependent errors are generated and the errors are averaged over time. Therefore, with DEM, the error becomes a noise-like signal and no harmonic distortion can be observed. However, due to randomization and distortion spreading, the noise floor is significantly increased [12, 23], as shown in Fig. 4.9. In short, DEM can only improve the SFDR/IMD, but not the SNR/SNDR. For applications requiring critical noise performance, such as in cellular base-stations, DEM is not suitable because of its high noise floor.

4.3 Design Techniques for Smart DACs

Though lots of design techniques have been developed to improve the intrinsic DAC's performance, most of them can only deal with global non-mismatch errors and systematic mismatch errors, which is absolutely not enough to achieve excellent performance after chip manufacturing, as because of process variations, the random mismatch errors are quite different from die to die. Other design techniques for intrinsic DACs that can deal with local random mismatch errors, such as DEM, which is blind to actual errors, can only decouple the distortion from the input signal so that no harmonic distortion is generated, but at the cost of trading noise performance with harmonic distortion performance as shown in Fig. 4.9.

As mentioned in Sect. 4.1, this work targets on designing a high-performance smart DAC, which is an intrinsic DAC with a calibration loop. By sensing and correcting the actual error, calibration techniques can be developed to cover both global non-mismatch and local mismatch errors, even together with the design techniques for intrinsic DACs which are already discussed in Sect. 4.2. Obviously, smart DACs have a significantly strong potential to gain better performance than intrinsic DACs.

As also mentioned in Sect. 4.1, depending on whether the error is corrected in the analog domain or in the digital domain, the calibration techniques for smart DACs are divided by analog calibration techniques and digital calibration techniques. In the following, existing analog and digital calibration techniques are discussed and compared. Based on the new concept of dynamic-INL introduced in Sect. 3.2, a novel digital calibration technique, called dynamic-mismatch mapping (DMM), is initially proposed in this section, and will be discussed in detail in next chapters.

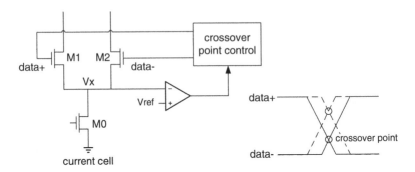

Fig. 4.10 Crossover-point control technique

4.3.1 Analog Calibration Techniques

4.3.1.1 Techniques for Non-mismatch Errors

Currently, for non-mismatch errors in DACs, such as sampling jitter, common duty-cycle error and finite output impedance, no related calibration techniques were published. To minimize the effect caused by these three non-mismatch errors, we still have to rely on the design techniques for intrinsic DACs, as discussed in Sect. 4.2.

To reduce the switching interference due to non-ideal cell switching which is described in Sect. 3.3.4, an analog calibration technique is proposed in [12] to control the crossover point of switching signals. As shown in Fig. 4.10, by monitoring the voltage bounce in Vx, the crossover point is controlled by a feedback loop, so that the bounce in Vx and related interferences during switching can be minimized.

4.3.1.2 Techniques for Mismatch Errors

Most of the analog design techniques for smart DACs focus on calibrating current sources, i.e. calibrating the amplitude error of current cells. Based on the literature investigation for published CMOS calibration DACs in recent 12 years, various silicon-proven analog calibration techniques for amplitude errors were investigated, as summarized in Table 4.2 [14–16, 18, 25, 30], with ascending order of their publishing date (earliest on the top).

As can be seen from Table 4.2, for the static accuracy of 14-bit and above, calibrating the amplitude error is necessary. Using current comparators (can be considered as 1-bit ADCs) and ADCs are two common methods to measure the amplitude error directly in time domain. For calibrating the amplitude error, the biasing voltage of the current source is trimmed using a floating-gate transistor [14, 18, 30] , as shown in Fig. 4.11a. In [15, 16, 25], a small calibration current-steering DAC (CAL-DAC) is used to calibrate the amplitude errors of current cells, as shown in Fig. 4.11b.

Table 4.2 Existing analog calibration techniques for amplitude errors

Reference	Existing analog calibration techniques		Static performance
	Error sensing method	Calibration method	
[18] ISSCC'00	$\Sigma\Delta$ ADC	Floating-gate trimming	0.5LSB@14b
[30] JSSC'03	Current comparator	Floating-gate trimming	0.3LSB@14b
[16] ISSCC'03	Off-chip 16b $\Sigma\Delta$ ADC	CAL-DAC	0.5LSB@14b
[15] ISSCC'03	Off-chip 6b SAR ADC	CAL-DAC	0.65LSB@16b
[14] ISSCC'04	Current comparator	Floating-gate trimming	0.65LSB@14b
[25] ESSCIRC'05	Current comparator	CAL-DAC	0.4LSB@12b

Regarding timing errors, up to this moment, no silicon-validated analog calibration techniques are published for DACs. Only some concepts are proposed in [54, 55]. In [54], by injecting a small current pulse, the glitch caused by the timing error can be compensated. However, this technique is still in concept phase and the circuits of measuring and calibrating the glitches are all missing, which could be a big challenge in the circuit design.

In [55], timing errors in the rising and falling edges are separately measured by a phase detector and are calibrated by tuning the substrate voltage of switch transistors in latches. However, at this moment, no silicon was designed to prove this concept yet.

4.3.2   Digital Calibration Techniques

4.3.2.1   Digital *vs.* Analog Calibration Techniques

As shown in the previous section, although the error-sensing circuits can be turned off when they are not used, analog calibration techniques still add additional active circuits into the DAC's core in order to correct mismatch errors. Those additional active circuits will create unwanted interference and add more parasitics. This is not desired for high-speed DACs that require the high-speed analog part to be small/clean/less parasitics as possible.

However, digital calibration techniques perform the correction in the digital domain, such as in the digital pre-processing block or in the decoder, instead of in the analog domain (i.e. in current cells). This is a big advantage of digital calibration techniques over analog calibration techniques. This means less overhead, less parasitics & less interferences in the analog core of DACs, and more room for smartness. As CMOS technology is continuously scaling down, digital circuits become more cost-efficient and powerful. Therefore, this advantage will become more and more attractive in the future. Moreover, digital calibration techniques are very easy to be ported to another technology node.

The fundamental difference between digital and analog calibration techniques is that digital calibration techniques improve the performance by reducing the effect of

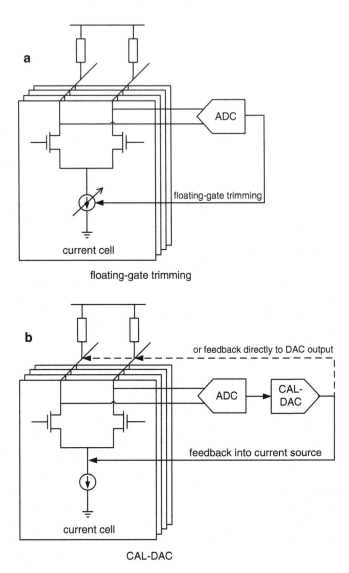

Fig. 4.11 Analog calibration techniques for the amplitude error

the error, while analog calibration techniques improve the performance by reducing the error itself. Although digital calibration techniques are quite different from analog calibration techniques, in most cases, they do not conflict with each other. In general, both digital and analog calibration techniques can be used together in one DAC. In this situation, digital calibration techniques add additional performance improvement on top of the performance achieved by analog calibration techniques.

In this section, existing digital calibration techniques for smart DACs are firstly discussed. Based on the analysis of the deficiencies in existing techniques, a novel digital calibration technique is proposed in this work to overcome the design challenges which are not covered by existing techniques. This new technique is initially introduced in this section and will be discussed in detail in Chap. 5.

4.3.2.2   Existing Digital Calibration Techniques: Mapping

If the nonlinear realtime transfer function of the DAC can be accurately measured and modeled, the distortion can be digitally compensated by adding an inverse transfer function as digital pre-processing before the DAC. However, measuring and modeling the nonlinear realtime transfer function of the DAC are very difficult, since the transfer function is typically signal- and frequency- dependent. In practice, regarding digital calibration techniques, currently only mapping techniques are developed and implemented for the DAC design.

Mapping concepts have been extensively developed to minimize the mismatch error in integrated circuits design. It defines a way of how to use existing circuit elements optimally based on the measured information. *For current-steering DACs, mapping means that the default switching sequence of the thermometer current cells can be mapped to a new switching sequence, in order to optimize the DAC performance.* The mapping operation is typically combined with normal binary-to-thermometer decoding operation. Since mapping techniques are based on the measured mismatch errors, the optimized switching sequence is different from sample to sample. Thus, a large freedom in mapping is required. The freedom of mapping depends on how the mapping decoder is implemented. For full mapping freedom, a reconfigurable memory is often used as a mapping decoder [35, 56, 57]. Since mapping is done in the decoder, it is fully digital. Therefore, it has all advantages belonging to digital calibration techniques.

According to how many kinds of errors are covered, mapping techniques can be categorized into one-dimensional and multi-dimensional mapping techniques. As the name suggests, one-dimensional mapping techniques only calibrate one specific kind of errors, while multi-dimensional mapping techniques calibrate multiple kinds of errors simultaneously. Obviously, multi-dimensional mapping techniques are more attractive and can have better performance than one-dimensional mapping techniques.

I. One-Dimensional Mapping

Similar to the situation of analog calibration techniques for smart DACs, all published one-dimensional mapping techniques with validated-silicon only focus on calibrating amplitude errors [35, 51, 56, 58], i.e. the main target is to improve the static performance (INL, DNL). Since amplitude errors are static errors, those mapping techniques belongs to static-mismatch mapping (SMM). Figure 4.12

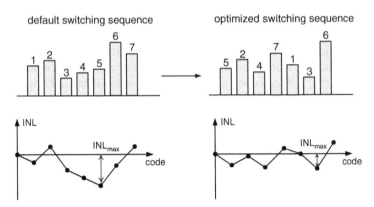

Fig. 4.12 Example of static-mismatch mapping (SMM)

Table 4.3 Summary of existing static-mismatch mapping techniques (SMM)

| Reference | Existing digital calibration techniques | | INL reduction factor |
	Error measurement	Correction method	
[35] JSSC'88	Current comparator	SMM (change switching sequence)	1.36
[56] TSCASI'05	VCO	SMM (change switching sequence)	1.78
[51]	Current comparator	SMM (change code distribution)	1.5
[58] JSSC'07	Current comparator	SMM (change switching sequence)	2.2

shows an example of static-mismatch mapping. As can be seen, after mapping, the INL is improved since the integral amplitude error is reduced. Table 4.3 summarizes existing static-mismatch mapping techniques for smart DACs. In [35, 56, 58], by measuring amplitude errors, the switching sequence of the thermometer current cells is optimized to reduce the INL based on amplitude-error cancelation. While in [51], benefit from parallel sub-DACs, similar to changing the switching sequence, the codes distributed to sub-DACs can be optimized to reduce both DNL and INL. As also shown, with static-mismatch mapping, the INL reduction factor is typically around 2, i.e. 1-bit improvement on the static linearity. Since the timing error is not covered by static-mismatch mapping (SMM), the improvement by SMM on the DAC's dynamic performance is only observed at very low frequencies and is negligible at high frequencies [51, 58].

Several concepts of one-dimensional mapping for timing errors have been proposed in [38, 59, 60]. Similar to static-mismatch mapping, mapping for timing errors optimizes the switching sequence by sorting timing errors such that the dynamic performance can be improved. As mentioned before, the DAC's dynamic performance is affected by both amplitude and timing errors. Just mapping for amplitude errors or just mapping for timing errors is not enough to guarantee a good performance at high frequencies. Therefore, multi-dimensional mapping techniques are required to calibrate both amplitude and timing errors at the same time.

II. Multi-dimensional Mapping

A concept for multi-dimensional mapping using cost function for the combination of amplitude and timing errors was firstly introduced in [61]. This concept requires both amplitude and timing errors to be separately measured. However, due to the limitation of the circuit implementation, it is very difficult to measure timing errors in sub-picosecond accuracy. In this work, a new method is proposed that combines amplitude and timing errors as vector errors in the frequency domain. Then, based on the measured vector errors, a novel multi-dimensional mapping technique called dynamic-mismatch mapping is introduced and validated with silicon results.

4.3.2.3 A Novel Multi-dimensional Mapping Technique: Dynamic-Mismatch Mapping

Though digital calibration techniques have lots of potential advantages as described previously in Sect. 4.3.2.1, currently existing digital calibration techniques, such as static-mismatch mapping, are only based on the static mismatch error (i.e. the amplitude error of current cells) and their main achievement is the improvement on the DAC's static performance (e.g. the INL). However, as explained in Sect. 3.2, the dominant source of mismatch errors in high-speed high-performance CS-DACs is the dynamic mismatch between current cells, not the static mismatch. This is because with increasing sampling frequency, the timing error becomes more and more important and dominant [2, 10, 12, 15], as proven in Sects. 3.2.1.2 and 3.2.1.3. Therefore, in order to achieve high dynamic performance for DACs with high sampling rates, the dynamic mismatch that includes both amplitude and timing errors has to be optimized, instead of just optimizing the static mismatch.

In this work, a novel digital calibration technique called **dynamic-mismatch mapping (DMM)** is introduced that is validated with experimental results in Chap. 7. As a multi-dimensional mapping technique, dynamic-mismatch mapping calibrates both amplitude and timing mismatch errors. It optimizes the switching sequence of thermometer current cells based on reducing the dynamic-INL (a parameter introduced in Sect. 3.2.3 based on combining amplitude and timing errors as vector errors), instead of reducing the static-INL in traditional static-mismatch mapping techniques. Thus, not only the DAC static performance, but also the dynamic performance can be improved. The details of dynamic-mismatch mapping technique will be given in Chap. 5. A dynamic-mismatch sensor based on a Zero-IF receiver has also been developed to measure the dynamic mismatch errors, as will be described in Chap. 6.

Since dynamic element matching (DEM) is also a digital performance-enhancement technique which can cover both amplitude and timing errors, it is necessary to compare dynamic-mismatch mapping (DMM) with DEM. Table 4.4 compares the proposed DMM calibration technique with existing digital calibration techniques, i.e. traditional static-mismatch mapping (SMM) and DEM. This work is the first reported DAC that measures both amplitude and timing mismatch errors,

Table 4.4 Comparison of digital calibration techniques for mismatch errors

Reference	Techniques	Targeted mismatch errors		Noise figure
		Amplitude error	Timing error	
[35, 51, 56, 57]	SMM	○		☺
[4, 23, 53]	DEM	◐	◐	☹
This work	DMM	●	●	☺

○ Only for systematic errors
◐ For both systematic and random errors, but these errors are not being measured. And noise performance is sacrificed
● Both systematic and random errors are measured

and corrects them in the digital domain. Moreover, due to the mismatch effect being reduced instead of randomized by DEM, the proposed DMM will not increase the noise floor. As will be verified by the experimental results of the proposed DMM DAC in Chap. 7, the proposed DMM DAC provides a comparable harmonic-distortion performance to the state-of-the-art DEM DACs [4, 23, 53], and a much better noise performance than those DEM DACs.

What has to be emphasized is that because mapping techniques improve the DAC's performance by only changing the switching sequence, the room of performance improvement in every chip sample depends on how good the default switching sequence is and how the mismatch error distributes. It's strongly recommended to design an intrinsic DAC as good as possible and have the mismatch errors linearly or Gaussian distributed, so that the errors can be easily canceled and mapping techniques can provide maximum benefits. In other words, mapping techniques are more suitable to achieve additional performance improvement on the top of an intrinsic DAC. If only relying on the design techniques for intrinsic DACs and analog calibration techniques for smart DACs, the improvement on the performance will become more and more difficult, or become less efficient. For example, in order to improve the performance of the state-of-the-art DACs, every extra bit in static accuracy or every additional 6 dB in the SFDR/IMD will significantly increase the complexity and cost in analog design. In that case, as mentioned in Sect. 4.3.2.1, digital calibration techniques, such as mapping techniques, can overcome this bottleneck and further improve the DAC's performance to the next level.

4.4 Summary of Design Techniques for Intrinsic and Smart DACs

In order to overcome the design challenges analyzed in Chap. 3, emerging silicon-proven design techniques for high-performance intrinsic and smart CS-DACs have been introduced in this chapter, as summarized in Table 4.5. In Table 4.5,

Table 4.5 Summary of emerging design techniques for high-performance intrinsic and smart DACs

	Technique	Reference	Non-mismatch error		Mismatch error	
			Output imp.	SWI	Amplitude	Timing
Intrinsic DACs	Always-on cascoding	[2, 48]	✓			
	Constant switching	[12, 49]		✓		
	Harmonic suppression	[50, 51]	✓	✓		
	CS layout techniques	[17, 19, 30–32, 40, 52]			○	
	Delay cancelation	[41]				○
	Dynamic element matching (DEM)	[4, 23, 53]			◑	◑
Smart DACs	Crossover control	[12]		✓		
	CS floating-gate trimming	[14, 18, 30]			●	
	CAL-DAC	[15, 16, 25]			●	
	Static-mismatch mapping (SMM)	[35, 51, 56, 57]			●	
	Dynamic-mismatch mapping (DMM)	This work			●	●

○ Only for systematic errors

◑ For both systematic and random errors, but these are not being measured. And noise performance is sacrificed

● Both systematic and random errors are measured

digital techniques for the DAC's performance enhancement are marked with gray background color.

For non-mismatch errors like sampling jitter and common duty-cycle error, currently we still rely on a good clock generator and intrinsic transistor/layout matching. For non-mismatch errors like finite output impedance and switching interference (SWI), always-on cascoding, constant switching, harmonic suppression and crossover control techniques are proposed to overcome these challenges, as discussed in Sects. 4.2.1 and 4.3.1.1.

For mismatch errors, most design techniques for intrinsic DACs, such as layout techniques for current sources and delay cancelation technique in [4, 10, 17, 19, 23, 30–32, 40, 41, 52, 53], can only deal with systematic amplitude or timing errors. Dynamic element matching (DEM) can reduce the harmonic distortion caused by both systematic and random amplitude or timing errors, however, at the cost of a significantly increased noise floor.

In order to further reduce mismatch errors, analog and digital calibration techniques have been developed for smart DACs to measure and calibrate the actual mismatch errors. Analog calibration techniques, such as floating-gate trimming in current sources and current compensation by using a lite calibration-DAC (CAL-DAC), are widely used in the published state-of-the-art DACs. Compared to analog calibration techniques, digital calibration techniques, such as traditional static-mismatch mapping (SMM) and the proposed dynamic-mismatch mapping (DMM), have less active circuits, less interference and less parasitics in the analog core of the DAC. Especially when pushing the DAC to extremely high performance, for analog calibration techniques, to increase one extra bit in the static accuracy or 6 dB in the SFDR/IMD/THD, will require substantial design effort including dedicated analog circuits and resources of power and area. In this case, due to the errors being corrected in the digital domain, digital calibration techniques will have less challenges and will be more efficient than analog calibration techniques.

Currently existing digital calibration techniques, i.e. static-mismatch mapping techniques (SMM) [35, 51, 56, 57], are only based on the amplitude error of current cells. As a result, the main benefit from those SMM techniques is the improvement on the DAC's static performance, such as the INL. However, the benefit on the DAC's dynamic performance is quite limited to low signal and sampling frequencies, because for high signal and sampling frequencies, the effects of timing errors becomes more and more visible. Therefore, in order to not only improve the DAC's static performance, but also improve the DAC's dynamic performance within the whole Nyquist band, both amplitude and timing mismatch errors have to be taken into account, especially for high sampling frequencies.

Dynamic-INL is a new parameter introduced in Chap. 3 to evaluate the dynamic mismatch between current cells, which covers both amplitude and timing mismatch errors. In contrast to static-mismatch mapping that optimizes the traditional static INL which is only based on amplitude errors of current cells, in this work, a novel digital calibration technique, called **dynamic-mismatch mapping (DMM)**, is introduced to optimize the dynamic-INL so that both the DAC's static and dynamic performance can be improved. This technique will be discussed in detail in Chap. 5.

Compared to DEM, as will be verified by experimental results of the proposed DMM DAC in Chap. 7, the proposed DMM DAC can provide a comparable harmonic-distortion performance to the state-of-the-art DEM DACs [4, 23, 53], and a much better noise performance than those DEM DACs.

4.5 Conclusions

In this chapter, design techniques for high-performance current-steering DACs have been discussed. For classifying these design techniques, the concept of smart DACs is introduced. A smart DAC is an intrinsic DAC with additional on-chip intelligence to enhance the performance/yield/reliability/flexibility. It always includes a feedback loop which is typically composed of information sensing circuits, processing circuits and actuators. Based on the measured actual chip information, every chip sample can achieve its best performance/reliability/life time. This big potential advantage of smart DACs over intrinsic DACs makes the development of design techniques for smart DACs highly necessary.

Based on where and how the errors are corrected, design techniques for smart DACs can be categorized by analog and digital calibration techniques. Comparison of analog and digital calibration techniques are given showing that digital calibration techniques have less overhead and less parasitics in the analog core of DACs. For a high speed design, a clean and small DAC core is preferred, which makes digital calibration techniques very attractive. Furthermore, in most cases, digital calibration techniques can be stacked on top of analog calibration techniques to achieve additional performance benefits.

Current digital calibration techniques for smart DACs only address on amplitude errors. As signal and sampling frequencies increasing, the effect of timing errors becomes more dominant than that of amplitude errors. In order to reduce the effect of both amplitude and timing errors, a novel digital calibration technique, called dynamic-mismatch mapping (DMM), is initially proposed In this chapter and will be discussed in detail in the next chapter. Compared to traditional static-mismatch mapping (SMM), DMM can improve the performance across the whole Nyquist band, especially at high frequencies. This advantage of DMM over SMM is due to both amplitude and timing errors being corrected. Compared to dynamic element matching (DEM), DMM does not increase the noise floor because the mismatch effect is reduced instead of randomized.

Chapter 5
A Novel Digital Calibration Technique: Dynamic-Mismatch Mapping (DMM)

In this chapter, a novel digital calibration technique, called dynamic-mismatch mapping (DMM), is proposed to correct the non-linear effect caused by both amplitude and timing mismatch errors. The theoretical background of this proposed DMM is firstly explained. How to implement DMM in an easy and efficient way is discussed next. Matlab Monte-Carlo statistical simulations are also performed in this section to show the performance improvement by DMM, with a comparison to traditional static-mismatch mapping techniques. Finally, the application of DMM is discussed and summarized.

5.1 Theory of Dynamic-Mismatch Mapping

As introduced in Sect. 4.3.2.2, mapping techniques for DACs are digital calibration techniques to optimize the switching sequence of current cells such that the DAC's performance can be improved. Figure 5.1 shows the Matlab simulation results of SFDR and THD of a 14-bit, 6Thermometer-8Binary (6T-8B) segmented DAC with five randomly chosen switching sequences of thermometer current cells (MSBs). As can be seen, with fixed amplitude and timing errors, the use of different switching sequences for MSBs in a very wide spread of performance. This implies that by just digitally optimizing the switching sequence of thermometer current cells (i.e. map the default switching sequence to an optimized switching sequence in the digital domain), both performance and yield can be significantly improved.

Existing static-mismatch mapping (SMM) techniques optimize the switching sequence only based on amplitude errors of current cells. However, in most applications, current cells are used as switched current cells, rather than static current sources. As defined in Sect. 3.2, dynamic mismatch is the mismatch between the dynamic switching behavior of current cells, including both amplitude and timing errors. An example of the dynamic-mismatch error in the differential output of the i-th current cell, in time, frequency and I-Q domain, are given in Fig. 5.2.

Y. Tang et al., *Dynamic-Mismatch Mapping for Digitally-Assisted DACs*, Analog Circuits and Signal Processing 92, DOI 10.1007/978-1-4614-1250-2_5,
© Springer Science+Business Media New York 2013

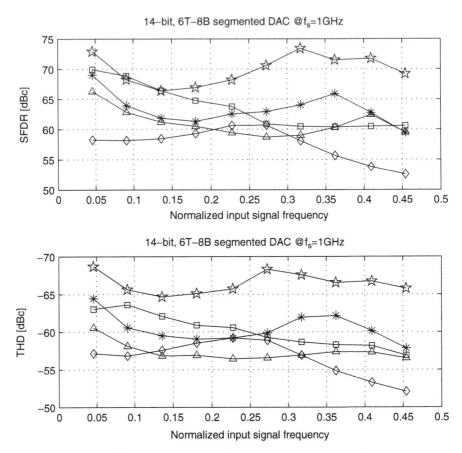

Fig. 5.1 SFDR and THD with five randomly chosen switching sequences of MSBs for the same DAC

Obviously, as concluded in Sect. 3.2.4, compared to static mismatch, dynamic mismatch represents the matching of switched current cells more completely and accurately. Similar to that the INL and DNL are developed to evaluate the static-matching between current cells, in Sect. 3.2, dynamic-INL and dynamic-DNL are also introduced to evaluate the dynamic-matching between current cells.

Instead of the traditional static-mismatch mapping optimizing the switching sequence and improving the DAC's performance by reducing the integral static-mismatch error (i.e. the INL) based on only amplitude errors [35, 51, 56, 57], the proposed dynamic-mismatch mapping (DMM) technique improves the DAC's performance by reducing the integral dynamic-mismatch error (i.e. the dynamic-INL) so that the error effect of both amplitude and timing errors can be reduced. The advantage of dynamic-mismatch mapping over static-mismatch mapping is that dynamic-mismatch mapping can improve the DAC's dynamic performance even at high frequencies, where a significantly less improvement can often be seen in

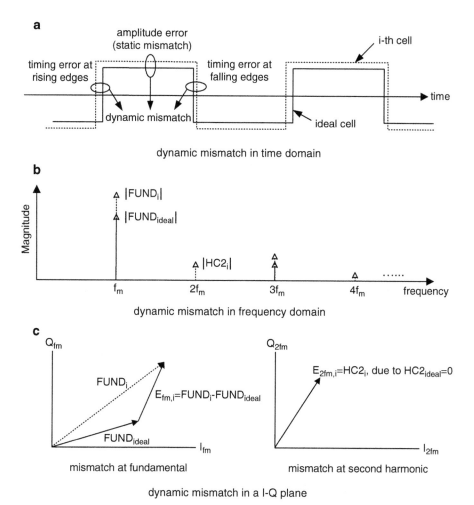

Fig. 5.2 Dynamic mismatch in time, frequency and I-Q domain

static-mismatch mapping in [35, 51, 56, 57]. This advantage will be verified by Monte-Carlo statistical analysis in Sect. 5.3 and silicon measurement results in Chap. 7.

As defined in Sect. 3.2.3, the dynamic-INL is given as:

$$
\text{dynamic-INL}_{code} = \sqrt{\frac{|E_{fm,code}|^2 + |E_{2fm,code}|^2}{|FUND_{ideal,LSB}|^2}}
$$

$$
= \sqrt{\frac{|\sum_i^{code} E_{fm,i}|^2 + |\sum_i^{code} E_{2fm,i}|^2}{|FUND_{ideal,1LSB}|^2}} \ \text{LSB}, \ code = 0 \sim \text{full scale}
$$

(5.1)

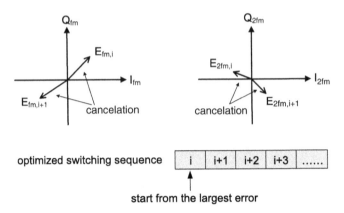

Fig. 5.3 Dynamic-mismatch mapping

The proposed dynamic-mismatch mapping (DMM) reduces the dynamic-INL according to certain criteria. A simple criterion is used in this work as an example: ***DMM changes the switching sequence of thermometer current cells, such that the dynamic-mismatch error of each cell can be maximally canceled by that of the following cell.*** More complex mapping criteria may reduce the dynamic-INL even more. However, this simple sorting criterion combines an easy implementation in hardware with also rather good performance.

Clearly, if we want to find two current cells whose dynamic-mismatch errors can cancel each other, since dynamic-mismatch errors are vector signals, they should be canceled in the I-Q plane. As shown in Fig. 5.3, the (i+1)-th current cell following the i-th current cell in the switching sequence is found by minimizing the summed dynamic-mismatch error of these two cells, i.e. find a cell from unsorted cells such that

$$|E_{fm,i} + E_{fm,i+1}|^2 + |E_{2fm,i} + E_{2fm,i+1}|^2$$
$$= (I_{fm,i} + I_{fm,i+1})^2 + (Q_{fm,i} + Q_{fm,i+1})^2 + (I_{2fm,i} + I_{2fm,i+1})^2$$
$$+ (Q_{2fm,i} + Q_{2fm,i+1})^2 \qquad (5.2)$$

is minimal; the (i+2)-th, (i+3)-th current cells in the switching sequence and so on are found one-by-one from the unsorted current cells by the same principle. The room for optimization reduces during mapping since the number of the unsorted cells become less and less. Therefore, the current cell with largest dynamic-mismatch error is compensated at the first place. By sorting all current cells in this way, the dynamic-INL can be reduced.

In summary, the proposed dynamic-mismatch mapping technique finds the optimized switching sequence based on the relative position between dynamic-mismatch errors of all current cells in the I-Q plane. The absolute values of the dynamic-mismatch errors have no effect on finding an optimal switching sequence.

Only the relative position between the dynamic-mismatch errors of all current cells in the I-Q plane is the deterministic factor in this digital mismatch-calibration technique. How much the dynamic-INL can be reduced and how much the DAC's static & dynamic performance can be improved by DMM will be discussed in Sect. 5.3.

5.2 Measurement of Dynamic-Mismatch Error

5.2.1 Measurement Flow

In order to optimize the switching sequence with dynamic-mismatch mapping, the dynamic-mismatch errors of all current cells have to be accurately measured. As proposed in Sect. 3.2.2, in contrast to static mismatch that can be easily measured in the time domain [14–16, 18, 25, 30], dynamic mismatch can be efficiently measured in the frequency domain by modulating current cells as rectangular-wave outputs. According to the definition of the dynamic-INL in (5.1), only the dynamic-mismatch errors at the fundamental and second harmonic (E_{fm} and E_{2fm}) need to be measured. Since E_{fm} and E_{2fm} are vector signals, both I and Q components of E_{fm} and E_{2fm} have to be measured. The easiest way to measure the I/Q component of a vector signal is using an I/Q mixer to convert both components of the signal to DC and then measure them. Based on this idea, a dynamic-mismatch error sensor based on a zero-IF receiver is developed in this work. The block architecture of this dynamic-mismatch sensor is shown in Fig. 5.4a, and the circuit implementation will be given in Chap. 6. Since an ideal current cell does not exist, the current cells are firstly measured relative to a reference cell (an arbitrary-chosen cell) as E'_{nfm}, then the actual dynamic-mismatch errors E_{nfm} of current cells can be achieved by subtracting the averaged errors. As shown, the output of the i-th current cell and of the reference current cell are modulated with opposite phases as rectangular waves, so that the ac signal at the summation node only comprises the dynamic-mismatch error ($E'_{i,nfm}$). Then, the mismatch signal is demodulated by a sine LO and down-converted to DC by a I/Q mixer. After converting $E'_{i,nfm}$ to $E_{i,nfm}$ by subtracting the averaged error, the I-Q components, such as $I_{fm,i}$ and $Q_{fm,i}$ for E_{fm}, will be digitized by ADCs. The LO can be chosen at f_m or $2f_m$ to measure E_{fm} or E_{2fm}.

As shown in Fig. 5.4b, the amplitude of the i-th current cell ($cell_i$) and of the reference cell ($cell_{ref}$) are A_i and A_{ref}, respectively. The duty-cycle of $cell_i$ and of $cell_{ref}$ are D_i and D_{ref}, respectively (note that in this work, the duty-cycle is defined as the absolute duration of the pulse width, not a ratio in percentage). The delay error of $cell_i$, relative to $cell_{ref}$, is Δt_i. The modulation frequency and period are f_m and T_m, respectively. Then, the modulated rectangular-wave outputs of $cell_i$ and $cell_{ref}$ can be expressed as:

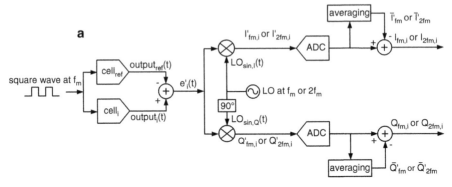

Measurement flow of dynamic-mismatch error, sine-wave LO

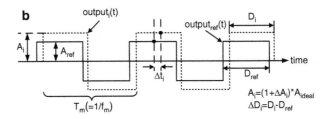

Modulated output of i-th and reference cells in time domain

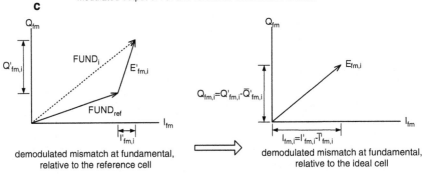

demodulated mismatch at fundamental,   demodulated mismatch at fundamental,
relative to the reference cell   relative to the ideal cell

Demodulated mismatch at fundamental in a I-Q plane

Fig. 5.4 I/Q demodulation by a sine-wave LO

$$output_{ref}(t) = \frac{4 A_{ref} D_{ref}}{T_m} \sum_{n=1}^{\infty} \frac{sin\frac{2\pi n f_m D_{ref}}{2}}{\frac{2\pi n f_m D_{ref}}{2}} sin(2\pi n f_m t)$$

$$output_i(t) = \frac{4 A_i D_i}{T_m} \sum_{n=1}^{\infty} \frac{sin\frac{2\pi n f_m D_i}{2}}{\frac{2\pi n f_m D_i}{2}} sin(2\pi n f_m t + n\frac{\Delta t_i}{T_m}2\pi) \qquad (5.3)$$

Next, the dynamic-mismatch error $e_i'(t)$ of $cell_i$, relative to $cell_{ref}$, can be derived as:

$$e_i'(t) = output_i(t) - output_{ref}(t) = \sum_{n=1}^{\infty} e_{nfm,i}'(t) \tag{5.4}$$

Since the LO is a sine-wave signal, only the mismatch-error at the fundamental frequency, i.e. $e_{fm,i}'$, is converted to DC. $e_{fm,i}'$ can be derived from e_i' as:

$$
\begin{aligned}
e_{fm,i}'(t) &= \frac{4A_i D_i}{T_m} \frac{sin\frac{2\pi f_m D_i}{2}}{\frac{2\pi f_m D_i}{2}} sin(2\pi f_m t + \frac{\Delta t_i}{T_m} 2\pi) \\
&\quad - \frac{4A_{ref} D_{ref}}{T_m} \frac{sin\frac{2\pi f_m D_{ref}}{2}}{\frac{2\pi f_m D_{ref}}{2}} sin(2\pi f_m t) \\
&= \frac{4A_i sin(\pi f_m D_i)}{\pi} sin(2\pi f_m t + \frac{\Delta t_i}{T_m} 2\pi) \\
&\quad - \frac{4A_{ref} sin(\pi f_m D_{ref})}{\pi} sin(2\pi f_m t) \tag{5.5}
\end{aligned}
$$

With reasonable mismatch value and modulation frequency, $sin(\pi f_m D_i) \approx 1$, hence $e_{fm,i}'$ can be simplified as:

$$e_{fm,i}'(t) \approx \frac{4A_i}{\pi} sin(2\pi f_m t + \frac{\Delta t_i}{T_m} 2\pi) - \frac{4A_{ref}}{\pi} sin(2\pi f_m t) \tag{5.6}$$

Then, the I-Q components $(I_{fm,i}', Q_{fm,i}')$ of the demodulated fundamental mismatch error $(E_{fm,i}')$ of $cell_i$, relative to $cell_{ref}$, can be calculated as:

I demodulation: $e_{fm,i}'(t) \times LO_{sine,I}(t) = e_{fm,i}'(t) \times A_{LO,sin} sin(2\pi f_m t - \phi_{LO})$

$$
\begin{aligned}
&\approx -\frac{2A_{LO,sin} A_i}{\pi} [cos(2\pi 2 f_m t + \frac{\Delta t_i}{T_m} 2\pi - \phi_{LO}) \\
&\quad - cos(\frac{\Delta t_i}{T_m} 2\pi + \phi_{LO})] \\
&\quad + \frac{2A_{LO,sin} A_{ref}}{\pi} [cos(2\pi 2 f_m t - \phi_{LO}) - cos(\phi_{LO})]
\end{aligned}
$$

$$\Rightarrow I_{fm,i}' \approx \frac{2A_{LO,sin}}{\pi} [A_i cos(\frac{\Delta t_i}{T_m} 2\pi + \phi_{LO}) - A_{ref} cos(\phi_{LO})]$$

Q demodulation: $e'_{fm,i}(t) \times LO_{sine,Q}(t) = e'_{fm,i}(t) \times A_{LO,sin}cos(2\pi f_m t - \phi_{LO})$

$$\approx \frac{2A_{LO,sin}A_i}{\pi}[sin(2\pi 2f_m t + \frac{\Delta t_i}{T_m}2\pi - \phi_{LO})$$

$$+ sin(\frac{\Delta t_i}{T_m}2\pi + \phi_{LO})]$$

$$- \frac{2A_{LO,sin}A_{ref}}{\pi}[sin(2\pi 2f_m t - \phi_{LO}) + sin(\phi_{LO})]$$

$$\Rightarrow Q'_{fm,i} \approx \frac{2A_{LO,sin}}{\pi}[A_i sin(\frac{\Delta t_i}{T_m}2\pi + \phi_{LO}) - A_{ref} sin(\phi_{LO})] \qquad (5.7)$$

where ϕ_{LO} is the phase difference between LO and the reference cell.

Finally, as shown in Fig. 5.4c, the actual mismatch error ($E_{fm,i}$) of $cell_i$ relative to the ideal cell can be derived by subtracting the averaged errors:

$$I_{fm,i} = I'_{fm,i} - \frac{1}{N}\sum_{n=1}^{N}I'_{fm,i}$$

$$Q_{fm,i} = Q'_{fm,i} - \frac{1}{N}\sum_{n=1}^{N}Q'_{fm,i}, \text{N=number of cells} \qquad (5.8)$$

The measured $E_{fm,i}$ with I-Q components ($I_{fm,i}$, $Q_{fm,i}$) in (5.8) can be used in dynamic-mismatch mapping to optimize the switching sequence. The same calculation can be applied when measuring $E_{2fm,i}$.

However, for practical circuit design, a square-wave LO is much easier to be generated than a sine-wave LO. With a square-wave LO, dynamic-mismatch errors at all other odd harmonics (e.g. $E'_{3fm,i}$, $E'_{5fm,i}$, etc.) will fold into DC together with the error ($E'_{fm,i}$) at the fundamental. As will shown in the next section, this does not change the relative positions between the dynamic-mismatch errors of the current cells and will not affect the mapping performance. Figure 5.5 shows the measurement flow with a square-wave LO demodulation. The square-wave LO is given as:

$$LO_{square,I}(t) = \frac{4A_{LO}}{\pi}\sum_{n=1}^{\infty}\frac{1}{2n-1}sin[2\pi(2n-1)f_m t - (2n-1)\phi_{LO}]$$

$$LO_{square,Q}(t) = \frac{4A_{LO}}{\pi}\sum_{n=1}^{\infty}\frac{1}{2n-1}cos[2\pi(2n-1)f_m t - (2n-1)\phi_{LO}] \qquad (5.9)$$

where A_{LO} is the peak amplitude of square wave and ϕ_{LO} is the phase difference between LO and the reference cell.

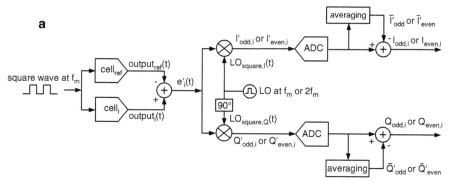

Measurement flow of dynamic-mismatch error, square-wave LO

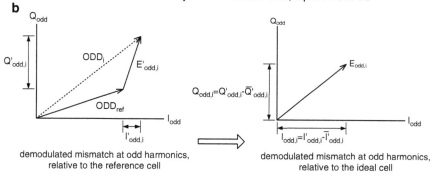

demodulated mismatch at odd harmonics,   demodulated mismatch at odd harmonics,
relative to the reference cell   relative to the ideal cell

Demodulated mismatch at odd harmonics in a I-Q plane

Fig. 5.5 I/Q demodulation by a square-wave LO

Then, the I-Q components $(I'_{odd,i}, Q'_{odd,i})$ of demodulated odd harmonics mismatch error $(E'_{odd,i})$, demodulated by the square-wave LO, can be derived as:

I demodulation: $e'_{odd,i}(t) \times LO_{square,I}(t) = \sum_{n=1}^{\infty} e'_{(2n-1)f_m,i} \times LO_{square,I}(t)$

$$\Rightarrow I'_{odd,i} \approx \frac{8A_{LO}}{\pi^2} \left(\sum_{n=1}^{\infty} \frac{A_i}{(2n-1)^2} cos[(2n-1)(\frac{\Delta t_i}{T_m}2\pi + \phi_{LO})] \right.$$

$$\left. - \sum_{n=1}^{\infty} \frac{A_{ref}}{(2n-1)^2} cos[(2n-1)\phi_{LO}] \right)$$

$$Q \text{ demodulation: } e'_{odd,i} \times LO_{square,Q}(t) = \sum_{n=1}^{\infty} e'_{(2n-1)f_m,i} \times LO_{square,Q}(t)$$

$$\Rightarrow Q'_{odd,i} \approx \frac{8A_{LO}}{\pi^2} \left(\sum_{n=1}^{\infty} \frac{A_i}{(2n-1)^2} sin[(2n-1)(\frac{\Delta t_i}{T_m} 2\pi + \phi_{LO})] \right.$$

$$\left. - \sum_{n=1}^{\infty} \frac{A_{ref}}{(2n-1)^2} sin[(2n-1)\phi_{LO}] \right) \tag{5.10}$$

According to the theory of Fourier series, $\sum_{n=1}^{\infty} \frac{1}{(2n-1)^2} cos[(2n-1)x] = \frac{\pi}{4}(\frac{\pi}{2} - x)$ for $0 \leq x \leq \pi$. Hence, $I_{odd,i}$ and $Q_{odd,i}$ can be rewritten as:

$$I'_{odd,i} \approx \begin{cases} \frac{2A_{LO}}{\pi}\{A_i[\frac{\pi}{2} - (\frac{\Delta t_i}{T_m} 2\pi + \phi_{LO})] - A_{ref}(\frac{\pi}{2} - \phi_{LO})\} \\ \quad \text{if } 0 \leq \phi_{LO} \leq \pi \\ \frac{2A_{LO}}{\pi}\{-A_i[\frac{3\pi}{2} - (\frac{\Delta t_i}{T_m} 2\pi + \phi_{LO})] + A_{ref}(\frac{3\pi}{2} - \phi_{LO})\} \\ \quad \text{if } \pi \leq \phi_{LO} \leq 2\pi \end{cases}$$

$$Q'_{odd,i} \approx \begin{cases} \frac{2A_{LO}}{\pi}\{A_i[\pi - (\frac{\Delta t_i}{T_m} 2\pi + \phi_{LO})] - A_{ref}(\pi - \phi_{LO})\} \\ \quad \text{if } -\frac{\pi}{2} \leq \phi_{LO} \leq \frac{\pi}{2} \\ \frac{2A_{LO}}{\pi}\{-A_i[2\pi - (\frac{\Delta t_i}{T_m} 2\pi + \phi_{LO})] + A_{ref}(2\pi - \phi_{LO})\} \\ \quad \text{if } \frac{\pi}{2} \leq \phi_{LO} \leq \frac{3\pi}{2} \end{cases} \tag{5.11}$$

Finally, as shown in Fig. 5.5b, the actual mismatch error ($E_{odd,i}$) of $cell_i$ relative to the ideal cell, demodulated by the square-wave LO at frequency f_m, can be derived by subtracting the averaged errors:

$$I_{odd,i} = I'_{odd,i} - \frac{1}{N} \sum_{n=1}^{N} I'_{odd,i}, \text{ and } Q_{odd,i} = Q'_{odd,i} - \frac{1}{N} \sum_{n=1}^{N} Q'_{odd,i} \tag{5.12}$$

where N is number of measured current cells. The same calculation can be applied when measuring the I-Q components ($I_{even,i}$, $Q_{even,i}$) of $E_{even,i}$ with a square-wave LO at $2f_m$ frequency. With square-wave demodulation, the dynamic-mismatch mapping will use $E_{odd,i}$ and $E_{even,i}$ to optimize the switching sequence, instead of $E_{fm,i}$ and $E_{2fm,i}$. As will shown in the next section, the same switching sequence will be found by dynamic-mismatch mapping, no matter if it is a sine-demodulation or square-demodulation.

5.2.2   Sine-Wave Demodulation vs. Square-Wave Demodulation

It is clear that, with sine-wave demodulation, the measurement results E_{fm} and E_{2fm} are demodulated from the fundamental and the second harmonic respectively,

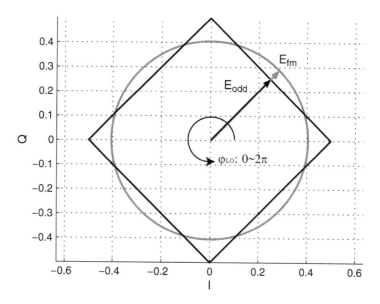

Fig. 5.6 Plots of E_{fm} and E_{odd} by sweeping ϕ_{LO}

while with square-wave demodulation, the measurement result E_{odd} and E_{even} are demodulated from odd and even harmonics respectively. As an example, Fig. 5.6 shows the difference between E_{fm} and E_{odd} for the same current cell. In this example, a sine-wave LO is assumed to be equal to the fundamental of a square-wave LO, i.e. $A_{LO,sin} = \frac{4}{\pi} A_{LO}$. As shown in Fig. 5.6, with LO phase (ϕ_{LO}) sweeping from 0 to 2π, the locus of the end points of E_{fm} and E_{odd} are a circle and a diamond respectively. Clearly, the absolute value of E_{fm} measured by a sine-wave LO is different from the absolute value of E_{odd} measured by a square-wave LO, and the difference depends on ϕ_{LO}. However, as mentioned in Sect. 5.1, dynamic-mismatch mapping is based on the relative positions of dynamic-mismatch errors of the current cells, not the absolute values. Therefore, if E_{odd} of all current cells have the same relative position as E_{fm} of all current cells, i.e. if the same optimized switching sequence can be found based on E_{fm} or E_{odd}, choosing sine- or square-demodulation will not affect the performance of dynamic-mismatch mapping.

Figure 5.7 shows an example of Matlab simulation results of E_{fm} and E_{odd} for ten current cells with different ϕ_{LO}. The first five current cells (1–5) have 0.1% amplitude error and (-2 ps, -1 ps, 0 ps, 1 ps 2 ps) delay error, relative to a reference cell. The second five current cells (6–10) have -0.1% amplitude error and (2 ps, 1 ps, 0 ps, -1 ps -2 ps) delay error, relative to the reference cell. The measurement frequency (f_m) is 50 MHz. As can be seen from Fig. 5.7, as the phase difference between LO and the reference cell (ϕ_{LO}) changes from 0 to 2π, the absolute values of E_{fm} and E_{odd} of all cells are different: the absolute positions of E_{fm} measured by a sin-wave LO are rotating with ϕ_{LO}, while the absolute positions of E_{odd} measured by a square-wave LO are rotating and parallelogram-like stretched with

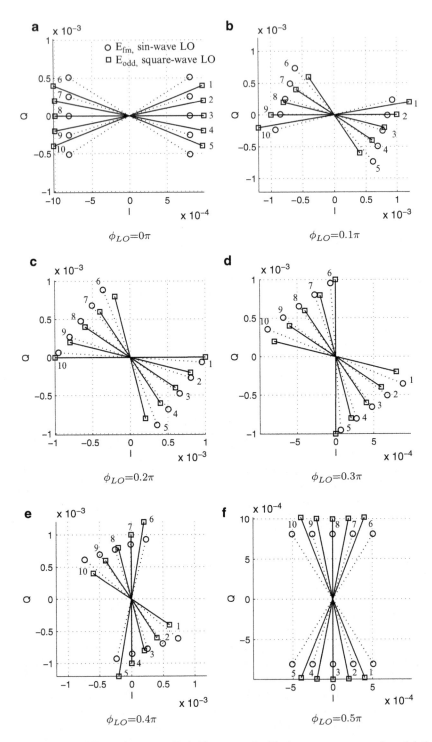

Fig. 5.7 E_{fm} and E_{odd} of current cells 1–10, measured with sine- or square-wave demodulation at different ϕ_{LO}

Table 5.1 Existing analog calibration techniques for amplitude errors

Cell number	Amplitude error (ΔA)	Timing error (delay error, Δt)
1	-0.1%	1 ps
2	-0.05%	-2 ps
3	0.05%	2 ps
4	0.15%	-3.5 ps
5	-0.15%	-3 ps
6	0.1%	-0.5 ps

ϕ_{LO}. However, the relative positions of E_{fm} of these ten current cells are the same as the relative positions of E_{odd} of these ten current cells, i.e. the most suitable current cell, whose dynamic-mismatch error can mostly cancel the dynamic-mismatch error of the previous cell, is the same for sine- or square-wave demodulation. The same conclusion applies to E_{2fm} and E_{even}. This means that with dynamic-mismatch mapping, the same optimal switching sequence will be found, no matter a sin- or square-wave LO is applied.

Since a square-wave LO is much easier to be generated than a sine-wave LO in circuit design, in this work, a square-wave LO is used to measure the dynamic-mismatch error information.

5.2.3   Weight Function Between Amplitude and Timing Errors

As discussed in the previous section, though both the shape (sine or square) and the phase of the LO have no influence on finding an optimized switching sequence, a different modulation frequency f_m will affect the relative positions of the dynamic-mismatch errors of all current cells and will lead to a different optimized switching sequence after DMM. This is because f_m changes the weight between amplitude errors and timing errors in the measurement results. This effect can be easily understood: for example, in the extreme case of $f_m = 0$ Hz where the dynamic-mismatch error only comprises amplitude errors, the switching sequence optimized by the proposed dynamic-mismatch mapping will be the same as that optimized by the traditional static-mismatch mapping.

As an example, Table 5.1 gives a group of six current cells with amplitude and timing errors. Figure 5.8 shows the DMM-optimized switching sequence based on different measurement frequencies ($f_m = 1$ MHz, 10 MHz, 50 MHz, 200 MHz, 1 GHz, respectively). The ϕ_{LO} doesn't affect the optimized switching sequence, so it is arbitrarily chosen to be 0.2π. As can be seen from Fig. 5.8a, at a very low measurement frequency, the DMM-optimized switching sequence is only determined by the order of the amplitude errors given in Table 5.1. As f_m increases, timing errors come in and become more and more dominant, resulting in different optimized switching sequences determined by the combined effect of amplitude and timing errors. When the measurement frequency is very high, as 1 GHz shown in

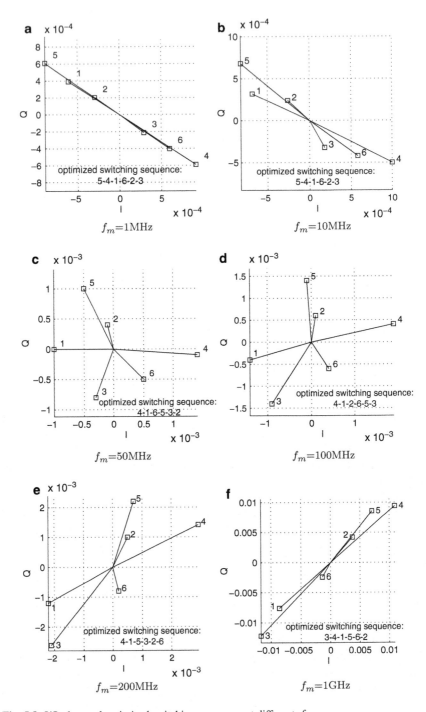

Fig. 5.8 I/Q plots and optimized switching sequences at different f_m

Fig. 5.8f, the DMM-optimized switching sequence is only determined by the order of timing errors given in Table 5.1. Clearly, it is f_m that determines the weight function between amplitude and timing errors.

In summary, **for a DAC with given mismatch errors, there is no universal best switching sequence; only a most suitable switching sequence exists for a defined application**. As mentioned in Sect. 3.2.4, the choice of f_m depends on applications, i.e. more weight on timing errors for high frequency applications requires a higher f_m and vice versa.

5.3 Theoretical Evaluation of DMM

In this section, Matlab Monte-Carlo statistical analysis is performed to evaluate the improvement on the DAC performance by the proposed dynamic-mismatch mapping (DMM). The evaluation process for DMM is shown in Fig. 5.9. Firstly, with given mismatch errors, the switching sequence is optimized by DMM based on the dynamic-mismatch errors measured at various modulation frequency f_m. The improvements by these optimized sequences on the DAC performance are analyzed to find the best f_m for the given mismatch errors and applications (sampling and signal frequencies), i.e. to find the best weight function between amplitude and timing errors so that an switching sequence most suitable for the given application can be found. Then, in order to check the robustness of DMM, the switching sequence optimized at this best f_m is checked for other applications (different sampling and signal frequencies) to show the performance improvement. Finally, the application of DMM and the comparison to other techniques are summarized.

A 14-bit 6T-8B segmented NRZ DAC is chosen as an example. Since the performance of a segmented DAC is typically dominated by the thermometer part (MSBs), the proposed DMM finds an optimized switching sequence of MSBs to reduce the dynamic-INL and improve the DAC's performance. The amplitude and timing mismatch errors are assumed to be zero-mean Gaussian distributed. The sigma (RMS) values for amplitude, delay and duty-cycle errors are 0.0125%, 5 ps and 5 ps, respectively.

5.3.1 Effect of f_m on Performance Improvement

As mentioned in Sect. 5.2.3, the modulation frequency f_m defines the weight function between amplitude and timing errors. Different weights between amplitude and timing errors results in different measurement results of dynamic-mismatch errors. As known, the proposed DMM reduces the dynamic-INL by optimizing the switching sequence of MSBs based on the measured dynamic-mismatch errors, so that the DAC's performance can be improved. Therefore, the performance

Fig. 5.9 Evaluation process for DMM

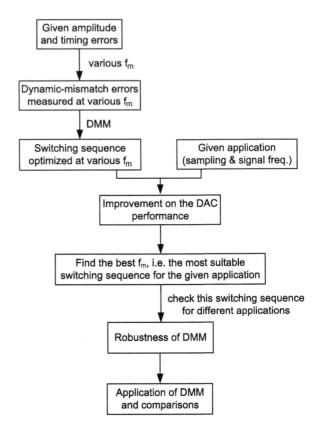

improvement by DMM is also dependent on f_m. In this section, the analysis of how f_m affects the DMM's improvement on the DAC's performance is given.

With the switching sequence being optimized by DMM at different f_m (chosen as 0 Hz, 5 MHz, 50 MHz, 500 MHz), the statistical Monte-Carlo simulation results of dynamic-INL, dynamic-DNL, SFDR and THD are summarized in Tables 5.2–5.5. The mean value ($\mu_{dynamic-INL}$, $\mu_{dynamic-DNL}$, μ_{SFDR}, μ_{THD}) of these simulation results are also plotted in Figs. 5.10 and 5.11. Several important phenomenons can be observed in Figs. 5.10 and 5.11:

- As discussed in Sect. 3.2.4, the dynamic-INL is a frequency-dependent parameter. When the modulation frequency f_m is 0 Hz (DC), it is equal to the static INL. As can been seen in Fig. 5.10a, the dynamic-INL without DMM at 0 Hz current-cell switching-rate, which is equal to the INL, is 2.1LSB. This is in line with the theoretical model given by (3.4): for 6-bit thermometer part with $\sigma_{amp} = 0.00125$, the INL is 0.0086LSB for the 6-bit thermometer part and 2.2LSB (0.0086LSB$*2^8$) for the total 14-bit accuracy. As current-cell switching-rate increases, the dynamic-INL increases, which means that the DAC's performance will decrease with sampling and signal frequencies.

Table 5.2 Dynamic-INL improvement by DMM with different f_m

f_m [Hz]	Dynamic-INL	w/o DMM	With DMM, the switching sequence is optimized at:			
			$f_m = 0$ Hz (SMM)	$f_m = 5$ MHz	$f_m = 50$ MHz	$f_m = 500$ MHz
0	$\mu_{dynamic-INL}$ $(= \mu_{INL})$	2.1LSB	0.94LSB	0.94LSB	1.1LSB	1.8LSB
	$\sigma_{dynamic-INL}$ $(= \sigma_{INL})$	0.7LSB	0.2LSB	0.2LSB	0.4LSB	0.6LSB
<30M	$\mu_{dynamic-INL}$	2.1LSB	0.94LSB	0.94LSB	1.1LSB	1.8LSB
	$\sigma_{dynamic-INL}$	0.7LSB	0.2LSB	0.2LSB	0.4LSB	0.6LSB
50M	$\mu_{dynamic-INL}$	2.3LSB	1.5LSB	1.2LSB	1.2LSB	1.9LSB
	$\sigma_{dynamic-INL}$	0.6LSB	0.4LSB	0.3LSB	0.3LSB	0.5LSB
63M	$\mu_{dynamic-INL}$	2.5LSB	1.8LSB	1.4LSB	1.3LSB	1.9LSB
	$\sigma_{dynamic-INL}$	0.6LSB	0.5LSB	0.4LSB	0.3LSB	0.5LSB
80M	$\mu_{dynamic-INL}$	2.7LSB	2.2LSB	1.6LSB	1.4LSB	2.0LSB
	$\sigma_{dynamic-INL}$	0.7LSB	0.6LSB	0.5LSB	0.3LSB	0.5LSB
100M	$\mu_{dynamic-INL}$	3.1LSB	2.8LSB	2.0LSB	1.5LSB	2.1LSB
	$\sigma_{dynamic-INL}$	0.8LSB	0.8LSB	0.6LSB	0.3LSB	0.5LSB
158M	$\mu_{dynamic-INL}$	4.5LSB	4.4LSB	3.1LSB	2.2LSB	2.5LSB
	$\sigma_{dynamic-INL}$	1.3LSB	1.3LSB	1.0LSB	0.4LSB	0.5LSB
200M	$\mu_{dynamic-INL}$	5.6LSB	5.5LSB	3.9LSB	2.7LSB	2.9LSB
	$\sigma_{dynamic-INL}$	1.7LSB	1.6LSB	1.2LSB	0.5LSB	0.6LSB

Table 5.3 Dynamic-DNL improvement by DMM with different f_m

f_m [Hz]	Dynamic-DNL	w/o DMM	With DMM, independent on switching sequences
0	$\mu_{dynamic-DNL}$ $(= \mu_{DNL})$	0.85LSB	0.85LSB
	$\sigma_{dynamic-DNL}$ $(= \sigma_{DNL})$	0.14LSB	0.14LSB
<30M	$\mu_{dynamic-DNL}$	0.85LSB	0.85LSB
	$\sigma_{dynamic-DNL}$	0.14LSB	0.14LSB
50M	$\mu_{dynamic-DNL}$	0.89LSB	0.89LSB
	$\sigma_{dynamic-DNL}$	0.14LSB	0.14LSB
63M	$\mu_{dynamic-DNL}$	0.92LSB	0.92LSB
	$\sigma_{dynamic-DNL}$	0.13LSB	0.13LSB
80M	$\mu_{dynamic-DNL}$	1.0LSB	1.0LSB
	$\sigma_{dynamic-DNL}$	0.12LSB	0.12LSB
100M	$\mu_{dynamic-DNL}$	1.1LSB	1.1LSB
	$\sigma_{dynamic-DNL}$	0.14LSB	0.14LSB
158M	$\mu_{dynamic-DNL}$	1.7LSB	1.7LSB
	$\sigma_{dynamic-DNL}$	0.23LSB	0.23LSB
200M	$\mu_{dynamic-DNL}$	2.1LSB	2.1LSB
	$\sigma_{dynamic-DNL}$	0.29LSB	0.29LSB

Table 5.4 SFDR improvement by DMM with different f_m

f_i [Hz]		SFDR w/o DMM	With DMM			
			$f_m = 0$ Hz (SMM)	$f_m = 5$ MHz	$f_m = 50$ MHz	$f_m = 500$ MHz
2.4 M	μ_{SFDR}	83 dB	93 dB	93 dB	92 dB	85 dB
	σ_{SFDR}	3.5 dB	2.2 dB	2.2 dB	3.6 dB	3.6 dB
15.1 M	μ_{SFDR}	79 dB	80 dB	84 dB	87 dB	84 dB
	σ_{SFDR}	2.1 dB	1.9 dB	2.1 dB	1.9 dB	3.2 dB
41.5 M	μ_{SFDR}	74 dB	74 dB	78 dB	82 dB	81 dB
	σ_{SFDR}	2.3 dB	2.8 dB	2.4 dB	2.4 dB	2.6 dB
83.5 M	μ_{SFDR}	71 dB	71 dB	76 dB	80 dB	80 dB
	σ_{SFDR}	3.1 dB	3.6 dB	3.1 dB	3.4 dB	2.8 dB
124.5 M	μ_{SFDR}	71 dB	71 dB	75 dB	80 dB	80 dB
	σ_{SFDR}	3.9 dB	4.3 dB	3.6 dB	3.5 dB	3.1 dB
166.5 M	μ_{SFDR}	72 dB	72 dB	76 dB	80 dB	79 dB
	σ_{SFDR}	3.7 dB	4.3 dB	3.6 dB	3.4 dB	2.6 dB
208.5 M	μ_{SFDR}	72 dB	72 dB	76 dB	80 dB	78 dB
	σ_{SFDR}	2.8 dB	3.1 dB	2.8 dB	2.8 dB	2.5 dB
247.6 M	μ_{SFDR}	71 dB	72 dB	75 dB	81 dB	79 dB
	σ_{SFDR}	3.7 dB	3.8 dB	3.7 dB	3.3 dB	3.4 dB

Table 5.5 THD improvement by DMM with different f_m

f_i [Hz]		THD w/o DMM	With DMM			
			$f_m = 0$ Hz (SMM)	$f_m = 5$ MHz	$f_m = 50$ MHz	$f_m = 500$ MHz
2.4 M	μ_{THD}	−79 dB	−86 dB	−87 dB	−87 dB	−81 dB
	σ_{THD}	2.3 dB	1.3 dB	1.4 dB	2.2 dB	2.5 dB
15.1 M	μ_{THD}	−73 dB	−73 dB	−77 dB	−80 dB	−78 dB
	σ_{THD}	1.7 dB	1.3 dB	1.4 dB	1.3 dB	1.8 dB
41.5 M	μ_{THD}	−70 dB	−70 dB	−73 dB	−77 dB	−76 dB
	σ_{THD}	1.7 dB	2.0 dB	1.9 dB	1.7 dB	1.9 dB
83.5 M	μ_{THD}	−68 dB	−68 dB	−72 dB	−77 dB	−76 dB
	σ_{THD}	2.3 dB	2.6 dB	2.4 dB	2.1 dB	1.9 dB
124.5 M	μ_{THD}	−68 dB	−68 dB	−72 dB	−76 dB	−76 dB
	σ_{THD}	2.5 dB	2.8 dB	2.4 dB	2.6 dB	2.1 dB
166.5 M	μ_{THD}	−68 dB	−68 dB	−72 dB	−76 dB	−75 dB
	σ_{THD}	2.4 dB	3.0 dB	2.5 dB	2.1 dB	1.8 dB
208.5 M	μ_{THD}	−68 dB	−68 dB	−72 dB	−75 dB	−74 dB
	σ_{THD}	2.0 dB	2.2 dB	2.0 dB	1.9 dB	1.9 dB
247.6 M	μ_{THD}	−68 dB	−68 dB	−71 dB	−76 dB	−75 dB
	σ_{THD}	2.4 dB	2.6 dB	2.6 dB	2.3 dB	2.2 dB

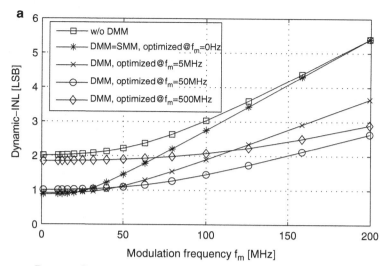

Dynamic-INL with the switching sequence optimized at different f_m

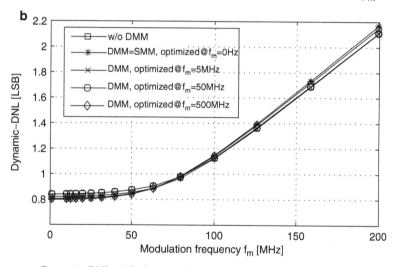

Dynamic-DNL with the switching sequence optimized at different f_m

Fig. 5.10 Dynamic-INL/dynamic-DNL improved by DMM with different f_m

- If the switching sequence is optimized by DMM with $f_m = 0$ Hz, the optimization is only based on amplitude errors and the proposed DMM becomes traditional SMM (static-mismatch mapping), resulting in an optimized dynamic-INL of 0.94LSB. The reduction factor is 2.2, i.e. roughly one bit benefit, which is a typical value in published SMM DACs [35, 51, 56, 58]. This one-bit improvement also gives around 9 dB improvement for SFDR and 6 dB for

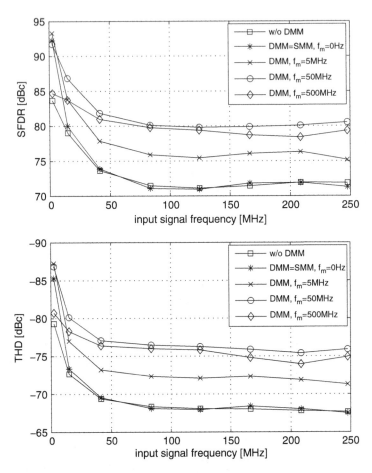

Fig. 5.11 SFDR/THD improvement by DMM with different f_m ($f_s = 500\,\text{MHz}$)

THD at very low signal frequencies, as shown in Fig. 5.11. However, as can be seen from Fig. 5.10a, this benefit decreases with the modulation frequency. For example, if the modulation frequency of the current cells are mostly at 200 MHz, there is no benefit by SMM or DMM with $f_m = 0\,\text{Hz}$. This expectation is verified by simulation results of SFDR/THD in Fig. 5.11. As can be seen, with a 500 MHz sampling frequency, at very low signal frequencies where the switching rate of the thermometer current cells are also very low, the SFDR/THD shows a 10 dB/7 dB improvement by SMM, while at high signal frequencies where the switching rate of the thermometer current cells are high, there is no improvement anymore. This observation is also often seen in silicon measurement results of published SMM DACs [51, 58]. This limitation further highlights the necessity of correcting the error effect of both amplitude and timing errors.

- With increasing f_m, the weight of timing errors in the measured dynamic-mismatch errors increases, and the weight of amplitude errors decreases. Thus, the switching sequence optimized by DMM tends to correct more on timing errors and less on amplitude errors. This means that the benefit from correcting amplitude errors is traded with the benefit from correcting timing errors. Fortunately, as can be seen from Fig. 5.10a, with f_m increased from 0 Hz to 50 MHz, a very little loss in the dynamic-INL improvement at low modulation frequencies returns a significant more improvement at high modulation frequencies. For example, for $f_m = 50$ MHz, the optimized dynamic-INL is 1.1LSB (0.16LSB less improvement than $f_m = 0$ Hz; the reduction factor is 1.9 compared to w/o DMM) at low modulation frequencies, however, the optimized dynamic-INL is 2.7LSB (2.8LSB more improvement than $f_m = 0$ Hz; the reduction factor is 2.1 compared to w/o DMM) at 200 MHz modulation frequency. Obviously, in this 14-bit DAC example, the DMM with $f_m = 50$ MHz is able to provide a 1-bit improvement for all modulation frequencies. This advantage will be more attractive at high sampling-rate DACs where the switching rate of the current cells is also high. For example, in this 500 MS/s DAC, for the SFDR/THD in Fig. 5.11, as f_m increases from 0 Hz to 50 MHz, the SFDR/THD is comparable at very low signal frequencies, but a significant improvement can be seen at high signal frequencies. The SFDR/THD optimized by DMM with $f_m = 50$ MHz is increased by at least 8 dB/7 dB at high frequencies.
- Further increasing f_m continuously decreases the weight on correcting amplitude errors and increases the weight on correcting timing errors. As shown in Fig. 5.10a, with $f_m = 500$ MHz, the DMM-optimized dynamic-INL almost equals the non-optimized dynamic-INL at low modulation frequencies, and the same is valid for the SFDR/THD at low signal frequencies. This is because f_m is so high and amplitude errors are very little corrected. However, at high modulation frequencies, the optimized dynamic-INL with $f_m = 500$ MHz shows a 2.4LSB improvement, similar to the case with $f_m = 50$ MHz. The same situation is applied to the SFDR/THD optimized by DMM with $f_m = 500$ MHz. This is because at 500 MS/s, the timing error is dominant at high signal frequencies and is well corrected. However, the amplitude error is not well corrected with $f_m = 500$ MHz, resulting in a slightly lower SFDR/THD than with $f_m = 50$ MHz. As a result, there is a weight balance between amplitude and timing errors. For a certain application (i.e. how high the sampling and signal frequencies are) and a certain DAC (i.e. how large its amplitude and timing errors are), f_m has to be properly chosen to maximize the benefit of DMM. For example, in this example of a 14-bit 500 MS/s DAC with 0.0125% amplitude error, 5 ps delay error and 5 ps duty-cycle error, f_m at 50 MHz is a good balance point. This is also the reason why we say there is no universal best switching sequence, but only a most suitable switching sequence exists for a defined application, as introduced in Sect. 5.2.3.
- Since the proposed DMM only changes the switching sequence of MSBs, it only reduces the effect of mismatch errors, not the actual mismatch errors.

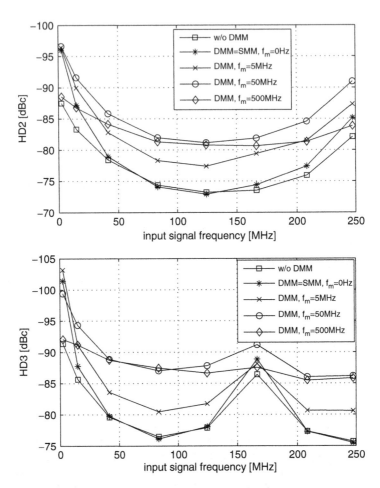

Fig. 5.12 HD2/HD3 improvement by DMM with different f_m ($f_s = 500$ MHz)

The dynamic-DNL is statistically the same with or without DMM, as shown in Fig. 5.10b. This is also the case for SMM, i.e. changing the switching sequence by mapping will not improve the DNL, which is confirmed by published SMM DACs in [56,58].

With the performance being improved, there is a corresponding improvement on the yield. In this example, since the SFDR/THD is statistically improved by at least 8 dB/7 dB across the whole Nyquist band, taking the 1-sigma spread of SFDR/THD shown in Tables 5.4 and 5.5 into consideration, this means at least 3σ improvement in the yield of SFDR/THD.

The Matlab statistical Monte-Carlo simulation results of the second/third harmonic distortion (HD2/HD3) of this DAC example are given in Fig. 5.12. The results are in line with the discussions above. Especially for the HD3, which is related to the

IM3 in multi-carrier narrow-band applications, an improvement of 10 dB is achieved by DMM at $f_m = 50\,\text{MHz}$. The peak of HD2 around $0.5\,f_s$ and the peak of HD3 around $0.33\,f_s$ are because the harmonics are approaching the sampling frequency f_s where the harmonics are strongly attenuated by the sinc function.

5.3.2 Robustness of DMM

As discussed in the previous section, for a given DAC, different applications, especially different sampling and signal frequencies, have different optimum switching sequences. If the maximum benefit from DMM is desired, dynamic-mismatch errors have to be measured at different f_m for different applications. However, in order to reduce the circuit design complexity, in a practical situation, we expect to have a fixed f_m to achieve a good enough performance improvement for most applications. By using the same 14-bit DAC example in the previous section, this section will check the robustness of the switching sequence optimized by DMM with $f_m = 50\,\text{MHz}$, to see how it behaves at different sampling frequencies.

As shown in Fig. 5.10a, with the switching sequence optimized by DMM at $f_m = 50\,\text{MHz}$, the optimized dynamic-INL shows an almost constant reduction factor of 2 for all current-cell switching-rates, compared to w/o DMM. This implies that even though the DAC has different sampling frequencies, the performance improvement by DMM will be almost the same. Figures 5.13 and 5.14 show the Matlab Monte-Carlo simulation results of SFDR and THD for this 14-bit DAC at different sampling frequencies ($f_s = 50\,\text{M}, 200\,\text{M}, 500\,\text{M}, 1\,\text{GHz}$). As can be seen, the improvement on SFDR/THD by DMM with $f_m = 50\,\text{MHz}$ is almost the same for different f_s. The simulation results show that in this DAC example, with $f_m = 50\,\text{MHz}$, both amplitude and timing errors are well calibrated, resulting in a constant performance improvement across the whole Nyquist band.

In summary, for a DAC with given mismatch errors, it is possible to find a fixed switching sequence that can improve the DAC performance for most applications. This fixed switching sequence may not be the best one for a specific application, but can be very close to it. In that sense, f_m should be properly chosen. The choice is dependent on the level of amplitude and timing mismatch errors which can be estimated in the design phase.

5.3.3 Application of DMM and Comparison to Other Techniques

As discussed, the proposed DMM minimizes the effect of mismatch errors by optimizing the switching sequence of thermometer current cells, but does not reduce mismatch errors. In other words, minimizing the effect of mismatch errors can

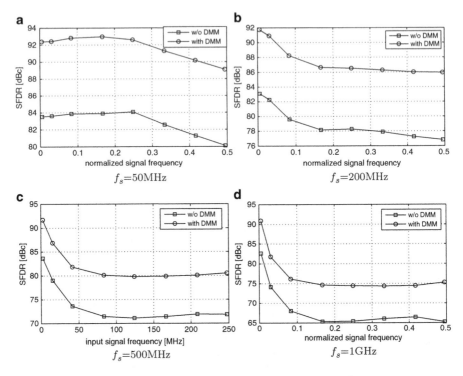

Fig. 5.13 SFDR improvement by DMM with $f_m = 50$ MHz at different f_s

be done in the digital domain, while reducing mismatch errors should be done in the analog domain. This is a fundamental difference between digital calibration techniques and analog calibration techniques, as discussed in Chap. 4.

Based on the statistical analysis in Sects. 5.3.1 and 5.3.2, with DMM, the DAC's static performance is typically improved by 1-bit, and the DAC's dynamic performance is typically improved by 7–10 dB. This improvement may not be interesting if we want to improve a DAC's dynamic performance from 60 dB to 70 dB, because this improvement can be easily done by good intrinsic circuit/layout design and analog calibration techniques. However, if we want to improve a DAC's dynamic performance from 70 dB to 80 dB even at hundreds MHz sampling frequency, the cost (area, power, design effort, etc.) of intrinsic circuit/layout design and analog calibration techniques becomes extremely high, especially because of timing errors. In such a situation, the proposed DMM is a very attractive "last-mile" solution, because as long as the DAC performance is dominated by mismatch errors, in theory, the DMM always provides almost 10 dB improvement in the whole Nyquist band based on a simple sorting algorithm, without increasing the noise floor and regardless the starting point is 70 dB or 90 dB. Another advantage of the proposed DMM is that it leaves most of algorithm and circuitry complexity in the

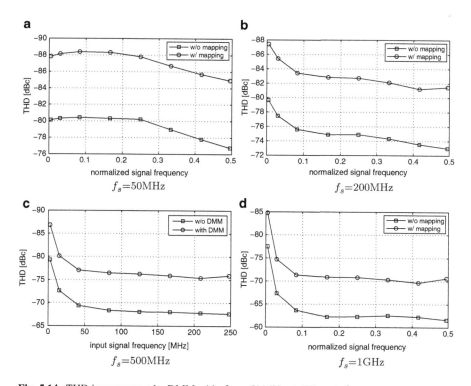

Fig. 5.14 THD improvement by DMM with $f_m = 50\,\mathrm{MHz}$ at different f_s

digital domain which will be more and more for free and powerful in the advanced CMOS technology, and keeps the DAC analog core clean.

Figure 5.15 shows a pyramid of design techniques that indicates how to achieve a DAC with ultra-good performance. The techniques for intrinsic and smart DACs are separated with different background colors, and the digital techniques are marked with*. As can be seen, the design techniques in level 1 and level 2 can be used together to achieve a rather good performance. However, if we want to push the performance further, traditional analog design techniques are not efficient anymore. In that sense, the digital techniques in level 3 can be stacked on top of the design techniques in level 1 and 2 to improve the performance further and efficiently, because they minimize the errors' effects in the digital domain and have minimal overhead of analog active circuits in the DAC core. Compared to static-mismatch mapping (SMM) and dynamic element matching (DEM), the advantages of the proposed DMM are significant:

- Compared to SMM that only improves the DAC's dynamic performance (e.g. SFDR, THD) at very low frequencies, DMM can provide a constant and significant improvement (around 10 dB in theory with a simple sorting algorithm) on the DAC's dynamic performance (e.g. SFDR, THD, HD2, HD3) across the

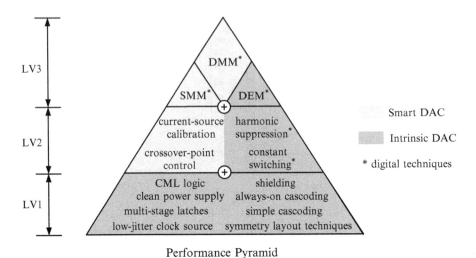

Fig. 5.15 Performance pyramid of design techniques

whole Nyquist band even at high sampling frequencies, as long as the mismatch errors are dominant.

• Compared to DEM, the proposed DMM does not increase the noise floor because the mismatch effect is reduced instead of randomized.

These advantages make the proposed DMM the best choice at level 3 of the performance pyramid shown in Fig. 5.15. It also suggests that a smart DAC can potentially achieve a better performance than an intrinsic DAC.

5.4 Conclusions

In this chapter, a novel digital calibration technique, called dynamic-mismatch mapping (DMM), has been introduced. The theoretical background of DMM is based on a new parameter, dynamic-INL, which is introduced in Chap. 3. Compared to the traditional static INL that can only predict the DAC's dynamic performance at very low frequencies, the proposed dynamic-INL can efficiently predict the effect of mismatch errors (including amplitude and timing errors) on the DAC's dynamic performance at all frequency bands. Instead of the traditional INL that is only based on the static amplitude errors, the proposed dynamic-INL is based on all mismatch errors and can be efficiently measured in the frequency domain. This introduces a new design methodology to evaluate the matching between current cells, which is more insightful than the traditional INL evaluated in the time domain.

By measuring the dynamic-mismatch errors in the frequency domain, an optimized switching sequence of the thermometer current cells is found by DMM to reduce the dynamic-INL, so that both the DAC's static and dynamic performance can be improved. In order to maximize the benefit from DMM, depending on the values of the amplitude and timing errors, a suitable measurement frequency (f_m) has to be chosen to balance the weight of calibration efforts between on amplitude errors and on timing errors. In general, more weights on timing errors for the applications with high sampling and signal frequencies requires a higher f_m, and vice versa.

Since both amplitude and timing errors are well corrected, compared to traditional current-source calibration techniques in [14–16,18,25,30] or static-mismatch mapping (SMM) in [35,51,56,58], as long as the DAC performance is dominated by mismatch errors, the proposed DMM is able to provide a constant and significant performance improvement (around 10 dB in theory with a simple sorting algorithm) across the whole Nyquist band, even at high sampling and signal frequencies, without increasing the noise floor and with minimal overhead in the DAC analog core. Compared to dynamic element matching (DEM), the proposed DMM does not increase the noise floor because the mismatch effect is reduced instead of randomized.

Chapter 6
An On-chip Dynamic-Mismatch Sensor Based on a Zero-IF Receiver

The proposed dynamic-mismatch mapping technique introduced in Chap. 5 requires the dynamic-mismatch errors of current cells to be accurately measured. The architecture of a novel on-chip dynamic-mismatch sensor based on a zero-IF receiver was already proposed in Chap. 5. In this chapter, the circuit design and performance analysis of the proposed dynamic-mismatch sensor are discussed.

6.1 Architecture Considerations

As described in Sect. 5.2, a zero-IF receiver based dynamic-mismatch sensor is an easy and efficient solution to the measurement of the mismatch between dynamic switching behavior of current cells. The implementation architecture of the proposed dynamic-mismatch sensor based on Fig. 5.5 is shown in Fig. 6.1. In stead of demodulating the I-Q at the same time by two paths as in Fig. 5.5, in this silicon implementation, only one demodulation path is implemented to save the circuitry. The I/Q demodulation is chosen by a I-Q phase selection in the input modulated signal. As shown, the I or Q component of the dynamic-mismatch error of $cell_i$, relative to the reference cell ($cell_{ref}$), is directly down-converted to DC and then measured.

The advantages of choosing a zero-IF or homodyne architecture are:

- Compared to a heterodyne architecture, the LO and the measurement frequency (f_m) can be easily generated. Both of them can be simply divided from the DAC's master sampling clock.
- Since a zero-IF receiver performs a direct, quadrature down-conversion on the wanted signal to the DC baseband, the design requirements on the filter and ADC are significantly relaxed.

The DC-offset voltage, a major disadvantage in zero-IF receivers, has no influence on the proposed dynamic-mismatch mapping (DMM) calibration technique.

Y. Tang et al., *Dynamic-Mismatch Mapping for Digitally-Assisted DACs*, Analog Circuits and Signal Processing 92, DOI 10.1007/978-1-4614-1250-2_6,
© Springer Science+Business Media New York 2013

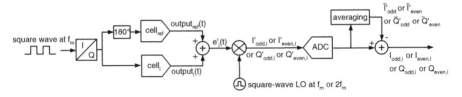

Fig. 6.1 Architecture of the proposed dynamic-mismatch sensor

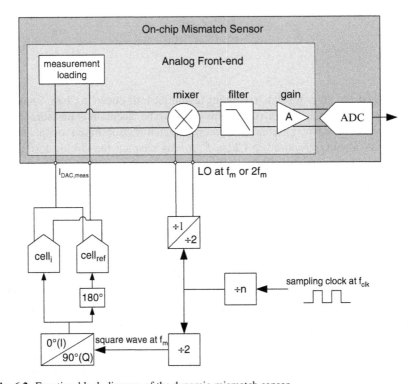

Fig. 6.2 Function-block diagram of the dynamic-mismatch sensor

It is explained in Chap. 5 that the dynamic-mismatch mapping only relies on the relative mismatches between current cells, not on absolute mismatch values.

A functional block diagram of the proposed on-chip dynamic-mismatch sensor is shown in Fig. 6.2. During the measurement, the current cell under measurement ($cell_i$) and the reference current cell ($cell_{ref}$) are switched with opposite phases as rectangular waves at the measurement frequency (f_m), which can be shifted digitally by 90° to measure I or Q, so that the AC part of the summed current output ($I_{DAC,meas}$) only comprises the relative mismatches between these two current cells. To simplify the circuitry, f_m can be chosen at $\frac{f_{clk}}{2n}$ (f_{clk}: master sampling clock, $n = 1, 2, 3, \ldots$), dependent on the desired weight function between amplitude and timing errors as mentioned in Chap. 5. The mismatch sensor includes an analog front-end to down-convert, filter and amplify the wanted signal, and an ADC to digitize the output of the analog front-end. The frequency of the LO can be chosen as f_m or $2f_m$ as required to measure the dynamic-mismatch errors.

Since the wanted signal is directly converted to DC, the noise in the low frequency band, especially the 1/f noise, has to be minimized. As will be described in the following sections, special care has been taken in circuit design to achieve a low-noise dynamic-mismatch sensor.

6.2   Analog Front-End Design

6.2.1   Circuit Design

6.2.1.1   Measurement Loading

Since the output of the current cell is current, the DAC is often terminated by a low impedance loading, e.g. the ac-equivalent differential $25\,\Omega$ as shown in Fig. 6.3. Therefore, the power of the mismatch signal will be too weak if it is directly measured at the normal output of the DAC. In order to increase the power of the mismatch signal, a second pair of cascode switches and a second output network are added to every current cell as shown in Fig. 6.3, so that a higher impedance loading can be connected. This additional output network is called measurement output network, with respect to the DAC's normal output network which is low-impedance terminated. This measurement output network is laid-out side-by-side

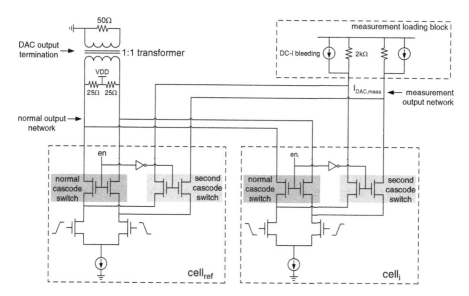

Fig. 6.3 Measurement output network and measurement loading

with the normal output network in order to have the same delay. All DAC cells have their own measurement-enable signal to choose which output network they will be connected to.

As shown in Fig. 6.3, the loading block at the measurement output network comprises a loading resistor and a DC-current bleeding source. The value of the loading resistor of the measurement output network is limited by the voltage headroom required for the normal operation of current cells. Since the mismatch signal is the AC part of $I_{DAC,meas}$, a DC-bleeding source takes 50% DC current of $I_{DAC,meas}$ away from the loading resistor so that even larger loading resistor can be used to increase the signal gain. For a 6T-8B segmented DAC with 20 mA full-scale output, in this 1.8 V dynamic-mismatch sensor design, the measurement loading resistor is chosen to be 2kΩ. Compared to differential 25Ω at the normal output network, the power of the mismatch signal is amplified by 44 dB with the proposed measurement loading block. Further increasing the current of the DC-bleeding source and loading resistor will give a large common-mode voltage variation at the measurement output network, when large current mismatches are present in current cells. If this cell-dependent common-mode voltage variation is large enough to affect the operation of current cells, a measurement inaccuracy will occur, especially for timing errors. Therefore, the measurement loading block has to be kept as a relatively DC low-impedance node.

To achieve the minimal mismatch between current-bleeding sources and DAC current cells, the bias current for the current-bleeding source is generated from the same current source array as DAC current cells.

6.2.1.2 Mixer

As mentioned before, to simplify the measurement circuitry, the wanted signal is directly down-converted to DC. Thus, a down-conversion Mixer with low 1/f noise is critical to the whole measurement sensitivity of the mismatch sensor.

A traditional Gilbert mixer, as shown in Fig. 6.4, has high 1/f noise, since the switching pair steers both the signal and the DC current to the load and the 1/f noise of the switching pair directly appears at the mixer's output without frequency conversion [62–64]. Passive mixers are widely used in zero-/low-IF receivers in communication systems due to the advantage of the low 1/f noise [64]. Compared to a Gilbert active mixer, a passive mixer contributes ultra low 1/f noise and high linearity because no DC current flows through the switching pair [64]. However, the disadvantage of the passive mixer is a signal loss of 3.9 dB, but this can be compensated by adding an amplifier stage after the passive mixer. A performance comparison between a Gilbert active mixer and a passive mixer is given in Table 6.1. Therefore, a passive mixer is more preferred in this design.

In this work, an AC-coupled current-driven passive mixer loaded with a low-pass filtering trans-impedance amplifier (TIA) is developed, as shown in Fig. 6.5. The passive mixer is loaded by a TIA with RC low-pass filtering to amplify the down-converted DC signal and filter out the high-frequency mixing products. Since the

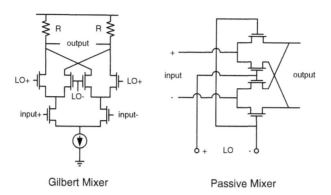

Fig. 6.4 Gilbert active mixer and passive mixer

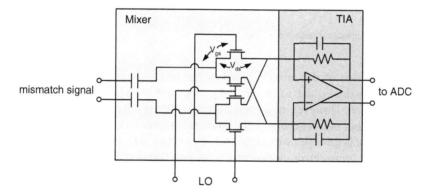

Fig. 6.5 Passive mixer terminated by a TIA

Table 6.1 Comparison between Gilbert and passive mixers	Gilbert mixer	Passive mixer
1/f noise	−	+
Linearity	−	+
Conversion gain	+	−
	$20 log_{10}(\frac{2}{\pi} g_m R)$dB	$20 log_{10}(\frac{2}{\pi}) = -3.9$ dB

TIA creates a low impedance node at the output of the mixer, i.e. a virtual ground. The linearity of this current-driven passive mixer is high, because the V_{ds} across the switches is very low so that the non-linearity of the switch's on-resistance is also low. Another source of the non-linearity of the on-resistance of the switch is the input-dependent V_{gs}. If the signal is DC-coupled to the mixer, since the common-mode voltage of the input signal is cell-dependent, the switch's on-resistance is also cell-dependent. This changes the signal transfer function of the analog front-end, resulting in measurement inaccuracy. The detailed analysis of the signal transfer function will be given in Sect. 6.2.2. In order to achieve the required measurement

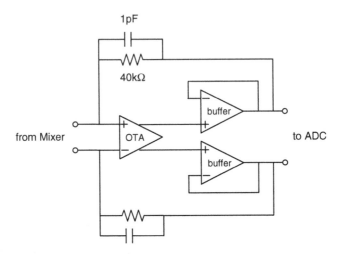

Fig. 6.6 Trans-impedance amplifier (TIA)

accuracy, the size of the mixer's switches has to be increased so that this non-
linearity effect due to the variation of V_{gs} does not change too much the signal
transfer function. However, larger switch transistors results in a larger capacitive
load at the output of LO generator and additional power consumption. And what is
more important is that even with large switch transistor size, it's still very difficult to
achieve sub-ps timing error measurement accuracy because the variation of V_{gs} still
can be very large with DC-coupling. In this design, an AC-coupling capacitor (1 pF,
metal-fringe cap) is used to couple the input signal to the mixer so that both V_{ds} and
V_{gs} are cell-independent. Compared to simply increasing the size of switches, it is a
more efficient solution.

6.2.1.3 Filter and Gain Stage: Trans-impedance Amplifier

As mentioned above, the filter and gain stages of this dynamic-mismatch sensor
are implemented by a trans-impedance amplifier (TIA) with low-pass filtering. As
shown in Fig. 6.6, the main gain stage of this TIA is an operational transconductance
amplifier (OTA). This OTA stage is followed by two buffers so that the low
impedance loading at the output will not decrease the gain of the OTA. The value
of the feedback resistor around the TIA is determined by the required mismatch-
measurement dynamic range, the output swing of the buffers and the input swing
of the ADC. In this design, a 40 kΩ N-poly resistor is used as the feedback resistor
to handle as large as 10 μA input current of the TIA within the swing limitation of
the buffer and the ADC. The feedback capacitor is a 1 pF metal-fringe cap which is
limited by the silicon area. As a result, this RC network provides a low-pass cut-off
frequency of 4 MHz to filter out the high-frequency mixing products from the output
of the Mixer.

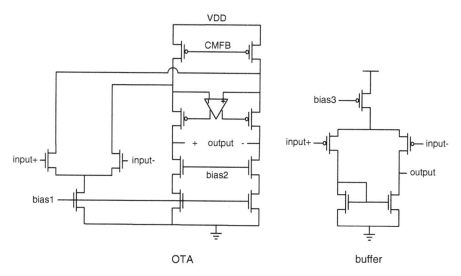

Fig. 6.7 OTA and buffer

The schematics of the OTA and the buffer are shown in Fig. 6.7. As the main gain stage of the TIA, the OTA has a folded gain-boosting topology. The buffer is an unity-feedback amplifier. The simulated DC closed-loop gain of the TIA with feedback loading is 81 dB. The unity-gain bandwidth of the closed loop is 184 MHz and the phase margin is 57°. The total power consumption of this TIA, including biasing, is 8 mW at 1.8 V supply. The noise performance of the analog front-end will be discussed later.

6.2.2 Signal Transfer Function

Figure 6.8 shows the complete circuit diagram of the dynamic-mismatch sensor. As shown, the dynamic-mismatch current (I_{signal}), i.e. the AC part of $I_{DAC,meas}$, is ac-coupled to the Mixer. $I_{ac\_coupled}$ is the effective current detected by the dynamic-mismatch sensor. Then, $I_{ac\_coupled}$ is mixed down to DC and amplified by the TIA. Thus, the wanted signal at the output of TIA is a DC voltage (V_o) which will be digitized by the ADC.

A single-ended circuit model of the AC-coupled signal path to the Mixer is shown in Fig. 6.9. C_{wire} is the parasitic capacitance of the measurement output network, and r_{on} is the on-resistance of the mixer switch. In this simplified circuit model, due to the high gain of the TIA, the input of the TIA can be considered as a virtual ground especially at low frequencies, i.e. the TIA has a zero input impedance ($Z_{in,TIA}$). It is obvious that a larger C_{ac}, a smaller r_{on} and a larger R_L will increase the signal gain. However, those parameters are limited by the operation condition of the current

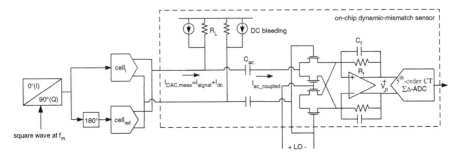

Fig. 6.8 Circuit diagram of the proposed dynamic-mismatch sensor

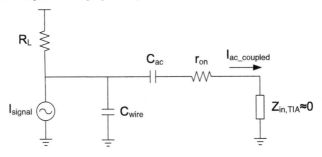

Fig. 6.9 Single-ended circuit model of the analog front-end before frequency translation

cell, silicon area and power. In this design, after performance-area-power trade-off, $C_{ac} = 1\,\text{pF}$, $r_{on} = 50\,\Omega$ and $R_L = 2\,\text{k}\Omega$ are chosen.

Based on the circuit model shown in Fig. 6.9 and assuming $Z_{in,TIA} = 0$, $I_{ac\_coupled}$ can be derived as:

$$I_{ac\_coupled} = I_{signal} \frac{j\omega R_L C_{ac}}{1 - \omega^2 R_L r_{on} C_{wire} C_{ac} + j\omega(r_{on} C_{ac} + R_L C_{wire} + R_L C_{ac})}$$

$$(6.1)$$

where $\omega = 2\pi f_m$, and f_m is the measurement frequency , i.e. LO frequency.

Equation (6.1) is consistent with Cadence Spectre periodic-state simulation (PSS) results of transistor-level circuits, as shown in Fig. 6.10. For example, at 50 MHz measurement frequency, together with $R_L = 2\text{k}\Omega$, $C_{wire} = 1.2\text{pF}$, $C_{ac} = 1\text{pF}$, $r_{on} = 50\,\Omega$, 37% of the mismatch current (I_{signal}) will be detected by this dynamic-mismatch sensor. For increasing measurement frequency, compared to the results calculated from (6.1), the signal gain from Spectre PSS simulation shows a small drop above 50 MHz (<3% up to 400 MHz). This is because of non-ideal LO transitions in circuit simulations. Some signal power is lost during the transition time of the LO, and the proportion of lost signal power over the total signal power increases with increasing LO frequency.

Finally, after $I_{ac\_coupled}$ is down-converted to DC by I or Q demodulation, the DC voltage (V_o) at the output of the TIA, which will be digitized by the ADC, can be derived as:

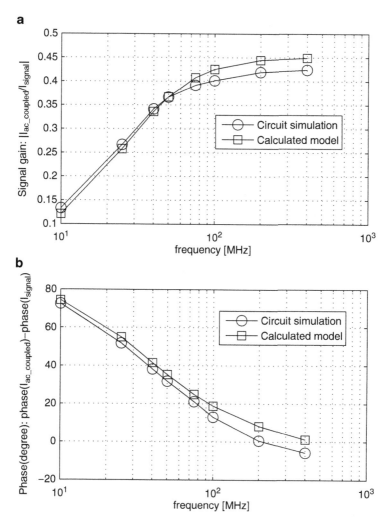

Fig. 6.10 Calculated and simulated signal transfer function (**a**) amplitude response versus frequency (**b**) phase response versus frequency

$$V_{o,I} = \frac{2}{\pi}|I_{ac\_coupled}|R_f sin(\phi_{LO})$$

$$V_{o,Q} = \frac{2}{\pi}|I_{ac\_coupled}|R_f cos(\phi_{LO})$$

(6.2)

where ϕ_{LO} is the phase difference between $I_{ac\_coupled}$ and the LO.

Figure 6.11 shows the plot of the simulated I/Q measurement results of the transistor-level analog front-end. In this circuit simulation, the timing error (delay error) of the DAC current cell is swept from -5 ps to 5 ps with 1 ps steps, and

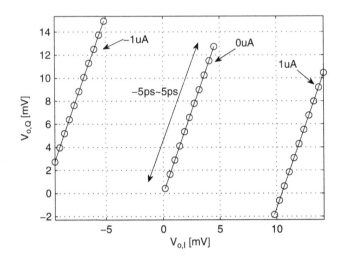

Fig. 6.11 Simulated I/Q measurement results of the analog front-end ($f_m = 50\,\text{MHz}$)

its amplitude error is swept from $-1\,\mu\text{A}$ to $1\,\mu\text{A}$ with $1\,\mu\text{A}$ steps. As can be seen, with a measurement frequency of $50\,\text{MHz}$, the sensitivity of this dynamic-mismatch sensor is $9.9\,\text{mV}/\mu\text{A}$ for amplitude errors and $1.3\,\text{mV/ps}$ for timing errors in this I-Q vector plane. As discussed in Sect. 5.2.2, the rotation angle of the I/Q plot is determined by the absolute phase difference between the signal and the LO. It has no influence on the dynamic-mismatch mapping because the rotation angle does not change the relative positions of the dynamic-mismatch errors. The centroid of the I/Q plot is not at zero due to the offset. However, the proposed dynamic-mismatch mapping technique only relies on the relative positions of the dynamic-mismatch errors, not absolute values. Therefore, any offsets also will not influence the performance of the dynamic-mismatch mapping.

6.2.3   Noise Analysis

In zero-IF receivers, noise can compromise the overall receiver sensitivity and most often the 1/f noise is the dominant noise source. In order to design for maximal measurement accuracy, the noise sources in the dynamic-mismatch sensor have to be analyzed. In this section, quantitative models to analyze the noise transfer function of the analog front-end are developed. The derived equations are verified by circuit simulations, and circuit optimizations to minimize the noise are also explained.

6.2.3.1 Noise Sources

Since there is a mixer in the analog front-end of the dynamic-mismatch sensor, the analog front-end is a time-varying circuit. Therefore, Cadence Spectre PSS and PNOISE simulations are performed to simulate the noise performance. Figure 6.12a shows the simulated noise power of the whole analog front-end and the first four dominant noise sources, integrated in a bandwidth of [10 Hz, 200 kHz]. The percentages of the noise from these noise sources over the total noise of the analog front-end are shown in Fig. 6.12b.

The noise from the DAC current sources and the measurement loading block has two parts: thermal noise and 1/f noise (also called flicker noise) . The 1/f noise from the DAC current sources and the measurement loading block are both modulated to the LO frequency after the mixer, which means it is far away from the signal band and will not affect the signal sensitivity. However, their thermal white noise at all harmonic frequencies of the LO are down-converted to the signal band and affects the sensitivity. As can be seen from Fig. 6.12a, the noise contributed by these thermal noise sources also increases with LO frequency. This is because the thermal noise present at the input of the mixer is high-passed due to AC-coupling. Therefore, a higher LO frequency will convert more non-attenuated thermal noise into the signal band. This thermal noise can only be minimized by narrowing the signal bandwidth, i.e. using a longer measurement time.

For the passive mixer, as mentioned before in Sect. 6.2.1.2, the 1/f noise is very low due to the fact that no DC current flows through this mixer. Confirmed by the circuit simulation result in Fig. 6.12a, no 1/f noise from the mixer can be observed. However, as seen, thermal noise of the mixer which is caused by the resistive channel of the MOS transistor still contributes to the noise in the signal band. Though the on-resistance of the switch is designed to be very low (50Ω in this design, limited by the area and LO driver strength), there are two magnification mechanisms to amplify the mixer's thermal noise. First, the thermal noise of the mixer is sampled at LO frequency due to switching. Due to aliasing, the thermal noise from the mixer in the signal band is significantly increased. Second, the thermal noise of the mixer is amplified by the TIA due to a so-called switched-capacitor effect. This noise amplification due to the switched-capacitor effect also operates on the noise from the OTA. As can be seen in Fig. 6.12a, both thermal noise of the Mixer and 1/f noise of the OTA are amplified by the same factor with increasing LO frequency. The amplification factor due to this switched-capacitor effect (SC effect) will be discussed in detail in the next section.

The thermal white noise of two $40\,k\Omega$ feedback resistors directly appears at the output of the analog front-end, and equals to $4kTR_f*2=3.11e\text{-}10V^2$ at $80\,^\circ C$. The thermal noise from these resistors does not change with LO frequencies as shown in Fig. 6.12a.

Table 6.2 summaries how noise sources contribute to the output noise of the analog front-end. As shown in Fig. 6.12b, due to the noiseamplification by the

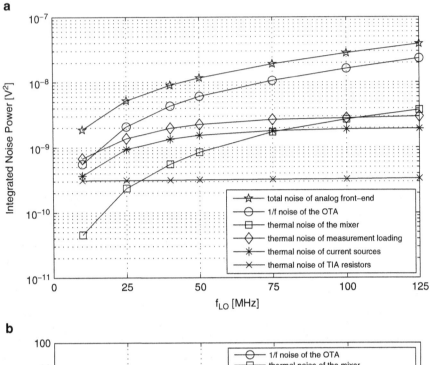

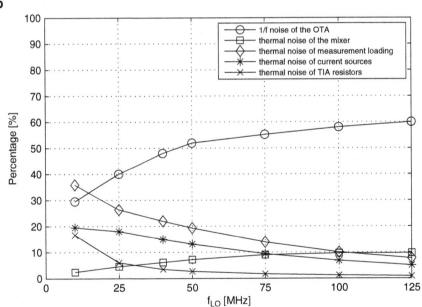

Fig. 6.12 Simulated noise performance and noise sources (**a**) simulated noise power of the analog front-end (**b**) percentage of the noise contribution of noise sources

Table 6.2 Transfer mechanism of noise sources to the output of the analog front-end

Noise source	1/f noise	Thermal noise
DAC current source itself	No, converted to f_{LO}	Yes, aliased due to mixing
Measurement loading block	No, converted to f_{LO}	Yes, aliased due to mixing
Passive mixer	No, too small	Yes, aliased due to sampling, and amplified by the SC effect
OTA	Yes, amplified by the SC effect	No, too small
TIA feedback resistor	–	Yes

SC effect, the 1/f noise from the OTA becomes the most dominant noise source at increasing LO frequency. A quantitative analysis of the noise amplification due to this SC effect will be given in the next section.

6.2.3.2   Noise Magnification Due to SC Effect

In order to analyze the noise magnification due to the switched-capacitor (SC) effect, a differential equivalent circuit of the analog front-end is shown in Fig. 6.13a. V_n, I_n are the input-referred noise voltage and current of the OTA, where $V_n = I_n Z_{in}$ (Z_{in}, the open-loop input impedance of the OTA). With impedance translation, the impedance network on the left side of the Mixer can be recomposed into the one shown in Fig. 6.13b. As can be seen, when the Mixer is switching, the capacitor C_{eq} is switched between the two inputs of the TIA. This generates a classic switched-capacitor equivalent resistor across the inputs of the TIA for each C_{eq}, which equals to $\frac{1}{f_{LO}C_{eq}}$ as shown in Fig. 6.13c. As a result, a low impedance path of R_{sc} is created, and the noise of the OTA will be amplified to the output of the TIA. Since the interesting frequency-band is at very low frequencies, C_f can be neglected. Then, the amplified OTA noise ($V_{n,out}$) at the output of the TIA due to this switched-capacitor effect can be derived as:

$$V_{n,out} = \left[\frac{\frac{R_{sc}}{R_f} + 1}{\frac{1}{A_v} + \frac{(1+\frac{1}{A_v})R_{sc}}{R_f} + \frac{R_{sc}}{A_v Z_{in}}} + \frac{\frac{R_{sc}}{R_{sc}Z_{in}}}{\frac{1}{A_v} + (R_{sc}||Z_{in})\frac{1+\frac{1}{A_v}}{R_f}} \right] V_n$$

when $Z_{in} \to \infty$ (i.e. normal case), $V_{n,out} = (\frac{R_f}{R_{sc}} + 1)V_n$ (6.3)

when $Z_{in} \to \infty$ & $R_{sc} \to 0$ (i.e. during short TIA inputs), $V_{n,out} = A_v V_n$

when $Z_{in} \to \infty$ & $R_{sc} \to \infty$ (i.e. during open TIA inputs), $V_{n,out} = V_n$ (6.4)

where A_v is the gain of the OTA. Then, the total noise amplification factor is contributed by the combination of these three cases. The contribution weight of each case is the ratio of its time slot over the LO period.

Previous work [64–66] on the analysis of the OTA noise amplification for similar passive Mixer architectures only considered the normal case of (6.4) and the input impedance network before the mixer was not taken into consideration. In

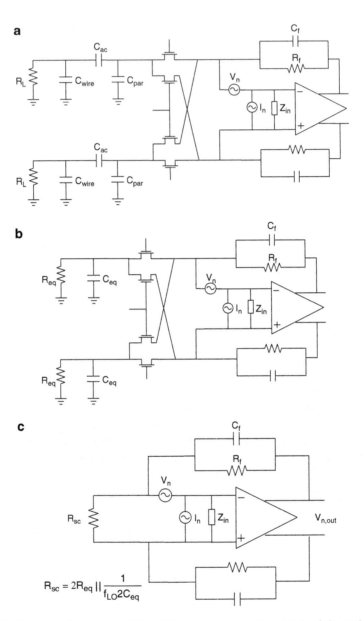

Fig. 6.13 Noise transfer analysis of the OTA noise (**a**) equivalent circuit of the analog front-end with OTA noise (**b**) equivalent circuit after impedance translation (**c**) equivalent circuit with switched-capacitor effect

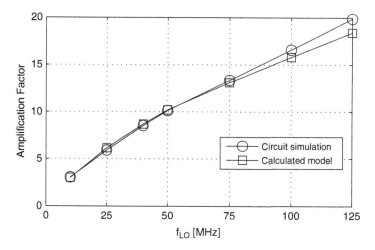

Fig. 6.14 Simulated and calculated noise amplification factor due to SC effect for OTA

this work, another two important circuit conditions, i.e. short and open TIA inputs, are also given in this noise transfer model. As seen, since the OTA noise could be significantly amplified, the OTA noise must be minimized, especially the $1/f$ noise. This noise minimization is done by circuit optimization, such as increasing the length of the input transistors of the OTA. In this design, after power-area-performance trade-off, the simulated integrated input-referred noise (V_n) of the OTA is 7.7 μV_{rms} in a frequency band of [10 Hz, 200 kHz].

Figure 6.14 shows the simulated amplification factor of the OTA noise due to the switched-capacitor effect. The simulation is based on Cadence Spectre PNOISE simulation on transistor-level circuits with a 100fF C_{par} at different LO frequencies (f_{LO}). The normal case in (6.4) is also plotted. As shown, the normal case in the derived (6.4) is well matched with the circuit simulation results at low LO frequencies. With increasing LO frequencies, the circuit simulated noise amplification factor is larger than that calculated, which is due to a larger noise amplification during the non-ideal switching transition of the mixer. During the time of the non-ideal switching transition, a lower impedance path is created between the inputs of the TIA, resulting in a larger noise amplification. This difference increases with LO frequencies because the weight of the non-ideal transition time in one LO period also increases with LO frequencies.

6.2.3.3 Non-overlap LO vs. Overlap LO

Similar to the amplification mechanism of the OTA noise caused by the switched-capacitor effect, if all Mixer's switches are ON due to an overlap LO, the OTA noise will be also amplified. This amplification factor during all switches are ON will be even higher than the case caused by the switched-capacitor effect because the

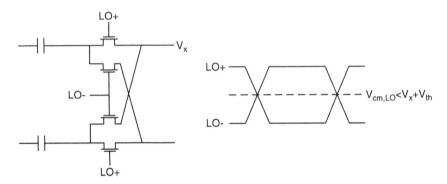

Fig. 6.15 Non-overlap LO

path created when all switches are ON most likely has a much lower impedance. Therefore, a non-overlap LO is required to avoid the possibility that all Mixer switches can be switched on at the same time. This is guaranteed by keeping the common-mode voltage ($V_{cm,LO}$) of the differential LO below the switch's turn-on threshold voltage shown in Fig. 6.15, so that all switches can never be ON at the same time. On the other hand, $V_{cm,LO}$ should not be too low because a LO with too low $V_{cm,LO}$ can turn off all Mixer switches during transition and decreases the signal gain as mentioned in Sect. 6.2.2. The transition of the LO should be made as fast as possible to have a low r_{on} so that the signal loss can be minimized.

6.3   ADC Design

Since a sigma-delta ADC is very suitable to measure a signal in low frequency bands, in this dynamic-mismatch sensor design, the same 5th-order continuous-time sigma-delta ADC with an integrated bandgap as in [67, 68] is used to digitize the output of the analog front-end. The reason to use a continuous-time (CT) $\Sigma\Delta$ ADC rather than a discrete-time (DT) $\Sigma\Delta$ ADC is that a CT $\Sigma\Delta$ ADC can act as an anti-alias filter and provide further low-pass filtering on the input signal. With a full-scale input of $1\,V_{rms}$ and a power consumption of 4.3 mW, this ADC has a measured performance of 92 dB SNR in a 200 kHz bandwidth [67,68], i.e. an integrated noise level of 25 μV_{rms}.

6.4   Overall Performance

The overall simulated performance of the proposed dynamic-mismatch sensor is summarized in Table 6.3. By defining the noise level as 0.5LSB of the measurement resolution, a measurement resolution of 22.4 nA for amplitude errors and 171fs for

Table 6.3 Simulated performance summary of the proposed dynamic-mismatch sensor

Technology	$0.14\mu m$ 1P6M baseline CMOS
Sensitivity@ $f_m = 50$ MHz	Amplitude error: 9.9 mV/μA
	Timing error: 1.3 mV/ps
Noise level@ $f_m = 50$ MHz	Analog front-end: 1.17e-8V^2, ADC: 6.25e-10V^2
Bandwidth=[10 Hz, 200 kHz]	Total: 1.2325e-8V^2 or $111\mu V_{rms}$
Resolution@ $f_m = 50$ MHz	Amplitude error: 22.4 nA
	Timing error: 171 fs
Power@1.8 V	Analog front-end: 10.7 mW, ADC: 4.3 mW
	Total: 15 mW
Area	Analog front-end: 0.08 mm^2, ADC: 0.19 mm^2
	Total: 0.27 mm^2

timing errors is achieved with a f_m of 50 MHz and a signal bandwidth of [10 Hz, 200 kHz]. Further improving measurement resolution requires a lower noise floor. This can be achieved by narrowing the signal bandwidth so that the noise, especially the thermal noise, can be reduced, but at the cost of a longer measurement time.

6.5 Conclusions

In this chapter, the circuit design of the proposed dynamic-mismatch sensor has been discussed. In order to achieve a low noise level with simple circuit architecture, a zero-IF receiver based dynamic-mismatch sensor is proposed. By analyzing the signal and noise transfer function, circuit models have been built to help circuit optimization. The derived models are confirmed by the transistor-level circuit simulations.

This sensor achieves a simulated measurement accuracy of 22.4 nA for amplitude errors and 171 fs for timing errors with a 50 MHz measurement frequency and a signal bandwidth of [10 Hz, 200 kHz]. Further improving measurement accuracy requires a lower noise floor. This can be achieved by narrowing the signal bandwidth so that the noise, especially the thermal noise, can be reduced, but at the cost of a longer measurement time.

Chapter 7
Design Example

In this chapter, a design example of a 14-bit $0.14\,\mu$m CMOS current-steering DAC with the proposed dynamic-mismatch mapping (DMM) is described. The intrinsic DAC core shows a performance of SFDR>65 dBc at 650 MS/s across the whole Nyquist band. The smart DAC with the proposed DMM achieves a performance of IM3<-83 dBc, SFDR>78 dBc and NSD<-163 dBm/Hz across the Nyquist band at 200 MS/s, which is at least 5 dB linearity improvement in the whole Nyquist band compared to the intrinsic performance, and the noise floor is not increased. Benchmarks are given, showing that this design has a state-of-the-art performance.

7.1 Overview

Figure 7.1 shows the architecture of the implemented 14-bit current-steering DAC. The DAC has a 6thermomter-8binary (6T-8B) segmented architecture: the six most significant bits (bit13~bit8) are implemented as thermometer current cells (63 MSB cells) and the eight least significant bits (bit7~bit0) are implemented as binary current cells. The DAC is implemented in a 1.8 V $0.14\,\mu$m CMOS baseline technology. The DAC can be configured in two modes: intrinsic-DAC mode and smart-DAC mode. The smart-DAC mode is the intrinsic DAC plus the proposed dynamic-mismatch mapping (DMM) technique introduced in Chap. 5. For the performance evaluation of the intrinsic DAC core, a standard binary-to-thermometer decoder is used and the dynamic-mismatch sensor is disabled. For the performance evaluation of the smart DAC with DMM, the dynamic-mismatch sensor is enabled and a mapping engine replaces the standard binary-to-thermometer decoder. The mapping engine integrates the function of the standard binary-to-thermometer decoder and the programming of the switching sequence of thermometer current cells. For flexibility, the sort logic for mapping is implemented off-chip in this prototype, but it is very easy to be integrated on-chip. The die photo of the 14-bit current-steering DAC with two modes is shown in Fig. 7.2. The circuit details will be described in the following sections.

Y. Tang et al., *Dynamic-Mismatch Mapping for Digitally-Assisted DACs*, Analog Circuits and Signal Processing 92, DOI 10.1007/978-1-4614-1250-2_7,

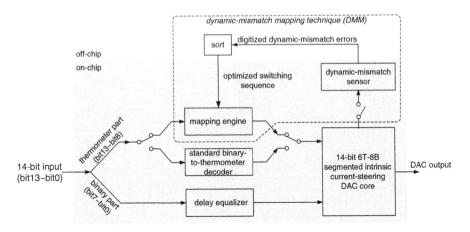

Fig. 7.1 Proposed DAC architecture with two modes

Fig. 7.2 Die photo

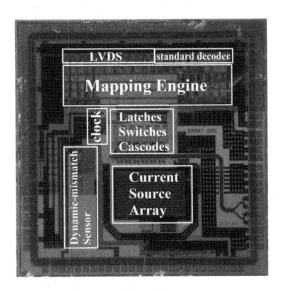

7.2 A 14-bit 650 MS/s Intrinsic DAC Core

7.2.1 Circuit Design

Figure 7.3 shows the block diagram of the 14-bit 650 MS/s intrinsic DAC core that is implemented in a 0.14 μm CMOS process. No calibrations or thick-oxide transistors are used in this intrinsic DAC. The design features are:

• The 14-bit current source array and two CML clocked-latch stages are reused from a 0.18 μm 12-bit DAC which is an improved version of [10]. Besides that,

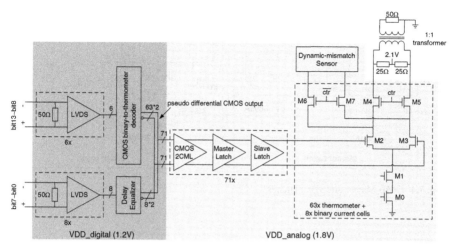

Fig. 7.3 Block diagram of the intrinsic DAC

an additional pair of cascode switches (M6, M7) is added to every thermometer current cell in order to measure its dynamic-mismatch errors. During the intrinsic-DAC mode, M6 &M7 are switched off and the dynamic-mismatch sensor is disabled. The circuit topology of the CML latches is shown in Fig. 7.4.

- For better signal integrity, a 50Ω-terminated LVDS interface is used to receive the 14-bit LVDS input data words. The circuit topology of the LVDS interface is shown in Fig. 7.5.
- Instead of using a CML decoder as in [10], a CMOS binary-to-thermometer decoder with a pseudo differential output is used in this design to save area and power consumption. Since this CMOS decoder is tool-synthesized, the design time and design effort are also significantly reduced. Different power supply domains are used to separate this CMOS decoder from the sensitive analog circuits, as shown in Fig. 7.3. Since the CML master and slave latches are used after the CMOS decoder to minimize the timing errors with minimum noise injection to the supply/substrate, a CMOS2CML converter converts the CMOS signal to the CML signal as the input of following CML master latch. The circuit topology of the CMOS2CML is shown in Fig. 7.5.
- Special care, such as dummies and tree-structure equal-length interconnection, has been taken in the layout to guarantee that each signal path has the same propagation delay.

This intrinsic DAC has a full-scale output current of 20 mA. As shown in Fig. 7.3, since the cascode switches (M4, M5) are thin-oxide transistors instead of thick-oxide transistors which are often used to allow a larger output voltage swing [2], this DAC is terminated with an effective differential impedance of 25Ω due to the output impedance and process over-voltage limitation. This results in a differential output voltage swing of $0.5V_{pp}$ or $0.177V_{rms}$. Therefore, assuming no transformer loss, the

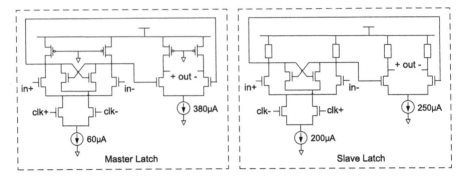

Fig. 7.4 CML Master and slave latches

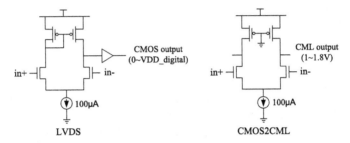

Fig. 7.5 LVDS interface and CMOS2CML converter

maximum power which can be delivered to the transformer's 50Ω load is -2 dBm. As explained in Sect. 2.4.3, if thick-oxide MOS are used as cascode switches (M4, M5) to tolerate a higher output swing, due to the current output of the DAC, a higher impedance loading can be connected to increase the power while keeping the same performance as long as the loading performs a linear I-V conversion and the nonlinear output impedance is not a dominant error source. The total power consumption is 260 mW at 650 MHz sampling frequency, with 1.2 V digital supply and 1.8 V analog supply. The active area of the intrinsic DAC is 1.1 mm².

7.2.2   Experimental Results

As mentioned above, since the current source array and latches of this 14-bit 0.14 μm intrinsic DAC are based on an improved version of a 12-bit 0.18 μm DAC in [10], in order to compare the performance, five samples were measured to evaluate this intrinsic DAC core and compare it with the previous 0.18 μm 12-bit DAC. Figures 7.6 and 7.7 show the measured THD (up to 11th harmonic)

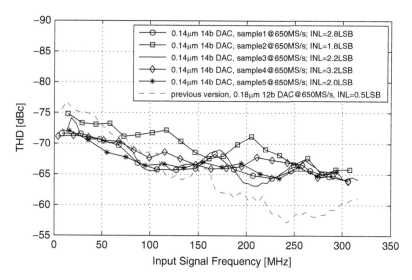

Fig. 7.6 Measured THD of the intrinsic DAC at 650 MS/s

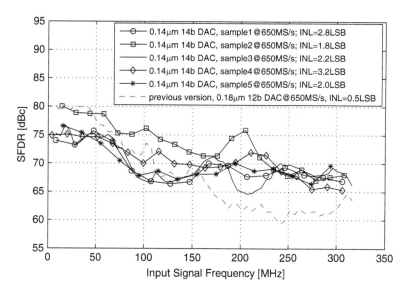

Fig. 7.7 Measured SFDR of the intrinsic DAC at 650 MS/s

and SFDR with full-scale output at 650 MS/s, respectively. This maximal sampling frequency is limited by the off-chip input-data generator (Agilent E4832A). The measured INL/DNL are 2.8 LSB/1.4 LSB, 1.8 LSB/2.2 LSB, 2.2 LSB/1.6 LSB, 3.2 LSB/1.9 LSB and 2 LSB/1.4 LSB for these five samples, respectively, based on the best-fit DC transfer curve.

Table 7.1 Performance summary of the 14b 650 MS/s intrinsic DAC core

Technology	0.14 μm 1P6M 1.8 V CMOS baseline[a]
Resolution	14-bit
Sampling rate	650 MHz
Full-scale output	20 mA, 1 $V_{pp.diff}$ for dc signal
(drive ability)	0.5 $V_{pp.diff}$ for ac signal (-2 dBm on 50Ω load)
avg. INL/DNL of 5 samples	2.4 LSB/1.7 LSB
SFDR, across whole Nyquist band	>65 dBc@650 MS/s
THD, across whole Nyquist band	<-63 dBc@650 MS/s
Power@650 MS/s	260 mW @ 1.2 V/1.8 V digital/analog supply
Active area	1.1 mm^2

[a]Only thin-oxide transistors are used

As shown, at 650 MS/s, this 14b intrinsic DAC core achieves THD <-63 dBc and SFDR>65 dBc across the whole 325 MHz Nyquist band. At signal frequencies below 25 MHz, the DAC linearity is constant with frequencies, which shows the linearity is dominated by static errors, such as amplitude error and finite output resistance. At signal frequencies above 25 MHz, timing errors start to dominate the linearity, resulting in an averaged 10 dB/decade roll-off for -THD and an averaged 15 dB/decade roll-off for SFDR. As also seen, this performance roll-off ends around 0.2 f_s (i.e. 130 MHz) and then the DAC linearity becomes flat. This phenomenon is due to the timing error effect and is confirmed by the theoretical analysis results in Sect. 3.2.1.2.

Compared to the previous version, which is a 12-bit 0.18 μm intrinsic DAC, this 14-bit 0.14 μm intrinsic DAC has similar SFDR/THD at low signal frequencies where static matching (i.e. the INL) is the dominant factor, but has around 5 dB improvement at high signal frequencies above 150 MHz. This improvement is partly due to the improved clock and signal interconnection so that the timing error is reduced, and partly due to the 2.1 V output termination voltage instead of 1.8 V so that the output impedance is improved. Table 7.1 summaries the performance of the 14b 650 MS/s intrinsic DAC core.

7.2.3   Comparison to Other Works

Figure 7.8 shows a comparison of the SFDR with published state-of-the-art DACs at high input signal frequencies (near Nyquist) shown in Table 2.1. As seen, this work, a 14b intrinsic DAC, achieves a performance of 65 dBc SFDR for a 325 MHz input signal frequency at 650 MS/s, which is close to those state-of-the-art DACs.

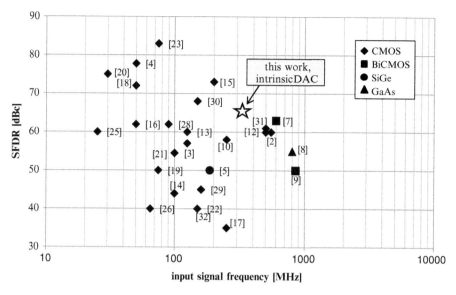

Fig. 7.8 SFDR of the intrinsic DAC core compared to state-of-the-art DACs

7.3   A 14-bit 200 MS/s Smart DAC with DMM

7.3.1   Circuit Design

The block diagram for the intrinsic DAC with the proposed dynamic-mismatch mapping (DMM) is shown in Fig. 7.9. Since the thermometer part is dominant in the DAC's performance, only the dynamic-mismatch errors of the thermometer current cells are measured. The additional pair of cascode switches (M6, M7) is enabled during measuring the dynamic-mismatch errors of current cells. The dynamic-mismatch errors of the MSB cells, relative to a reference cell (an arbitrary MSB cell), are measured off-line one-by-one by the dynamic-mismatch sensor as introduced in Chap. 6. The measurement frequency (f_m) is 45 MHz. An optimized switching sequence of the MSB cells is achieved by sorting the measured dynamic-mismatch errors so that the dynamic-INL can be reduced, as explained in Chap. 5. For flexibility, this sorting logic is implemented off-chip in this prototype, but it is very easy to be integrated on-chip. This optimized switching sequence with $f_m = 45$ MHz will be used to evaluate the performance of the DMM technique.

The memory-based mapping engine, highlighted in Fig. 7.10, is based on a register file with a size of 64rows*63columns. The row decoder selects a row of the memory as the decoded output, according to the DAC's input word. The switching sequence of 63 MSB cells is programmed and preset through the write port of the memory, so that each row corresponds to the output state of the 63 MSB current cells for the corresponding DAC input. In order to increase the speed, four

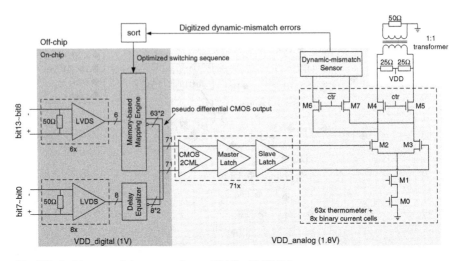

Fig. 7.9 Architecture of the proposed smart DAC with DMM

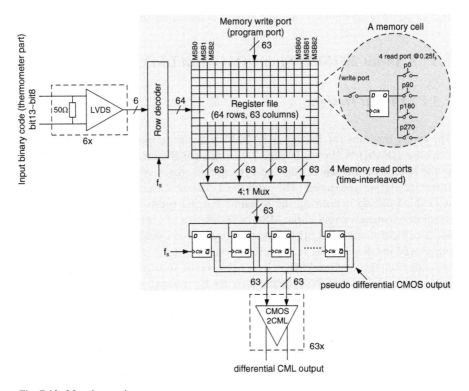

Fig. 7.10 Mapping engine

time-interleaved reading ports operating at a quarter of the DAC sampling frequency ($0.25\,f_s$) are used to read the memory. A 4:1 MUX array combines these four 63-bit outputs at $0.25\,f_s$ into a 63-bit output stream at f_s. This stream is synchronized by a DFF stage and is converted to pseudo differential CMOS signals. A CMOS2CML converter shown in Fig. 7.9 converts the rail-to-rail pseudo differential CMOS signal to a real low-swing CML differential signal. Then two CML latch stages are used to minimize the timing errors as shown in Fig. 7.9, with minimum noise injection into supply and substrate.

The proposed DMM DAC, together with the dynamic-mismatch sensor, has an active area of $2.4\,\text{mm}^2$ and consumes 270 mW at 200 MS/s with 1 V digital supply and 1.8 V analog supply.

7.3.2   Experimental Results

7.3.2.1   Improvement on Static Performance

As introduced in Chap. 5, calibration techniques for DACs can be categorized into analog calibration techniques and digital calibration techniques. Analog calibration techniques improve the DAC's static performance by calibrating or trimming current sources, while digital calibration techniques improve the static performance only by digital pre-processing. The proposed dynamic-mismatch mapping (DMM) belongs to the digital calibration techniques. Therefore, in this section, the comparison to other digital calibration techniques, i.e. traditional static-mismatch mapping (SMM) techniques [35,51,56,57], is given.

For static-performance measurement, the transformer shown in Fig. 7.9 is removed. Therefore, with 20 mA full-scale current output, the full-scale output voltage for DC signal is $1\,V_{pp,diff}$. The DAC's static performance is dominated by the thermometer current cells (MSB cells). Figure 7.11 shows the measured INL and DNL for the MSB cells, without mapping technique, with traditional static-mismatch mapping (SMM; the optimized switching sequence is only based on measured amplitude errors) and with the proposed dynamic-mismatch mapping (DMM; the optimized switching sequence is based on dynamic-mismatch errors measured at 45 MHz), respectively. As shown, with the same sorting algorithm, the INL is improved from 3.2 LSB to 1.7 LSB with SMM and to 1.8 LSB with DMM. The INL reduction factor is 1.78 for DMM, which is close to the theoretical 1-bit improvement given in Sect. 5.3.3. Compared to the traditional SMM, the proposed DMM has a negligible less improvement (0.1 LSB) on the DAC's static linearity (the INL), but as shown soon, it gains significant more improvement on the DAC's dynamic linearity (e.g. IM3, SFDR, THD), especially at high frequencies.

As explained in Chap. 5, since mapping techniques only change the switching sequence, the DNL will not be improved by mapping techniques. As shown in Fig. 7.11, the DNL before and after mapping are all about 2 LSB.

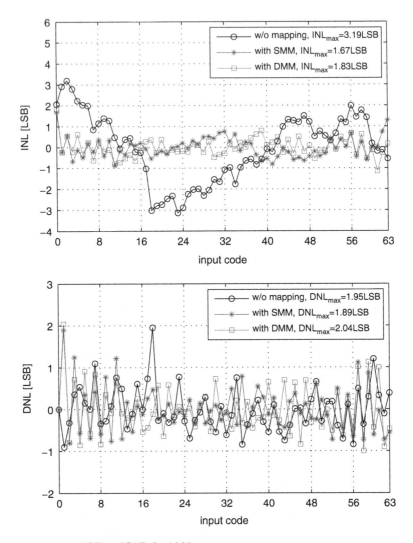

Fig. 7.11 Measured INL and DNL for 14-bit accuracy

7.3.2.2 Improvement on Dynamic Performance

Figure 7.12 shows the measured IM3 and noise power spectral density (NSD), without mapping, with traditional static-mismatch mapping (SMM) and proposed dynamic-mismatch mapping (DMM), respectively. The sampling frequency is 200 MHz which is limited by the switching interference from the mapping engine. As shown, with SMM the improvement on IM3 reduces gradually with signal frequencies, which means that the benefit from only correcting amplitude errors decreases and is almost negligible above 90 MHz. However, the proposed DMM

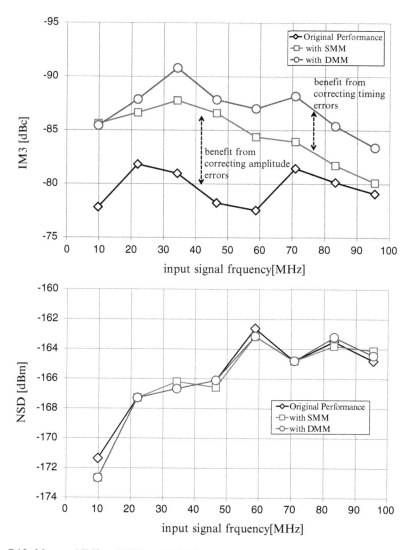

Fig. 7.12 Measured IM3 and NSD at 200 MS/s

provides an additional benefit on IM3 by also correcting timing errors, especially at high frequencies, resulting in a total improvement of 10 dB at low frequencies and still 5 dB up to Nyquist frequency. As seen, compared to SMM, the improvement from DMM by correcting timing errors increases with frequencies. At frequencies above 70 MHz, the improvement by DMM is limited by the finite output impedance, which is confirmed by the 40 dB/decade roll-off in the IM3 plot. Unlike dynamic-element matching (DEM), DMM does not increase the noise floor because the mismatch effect is reduced instead of randomized. The NSD remains <-163 dBm/Hz, independent of mapping, as shown in Fig. 7.12.

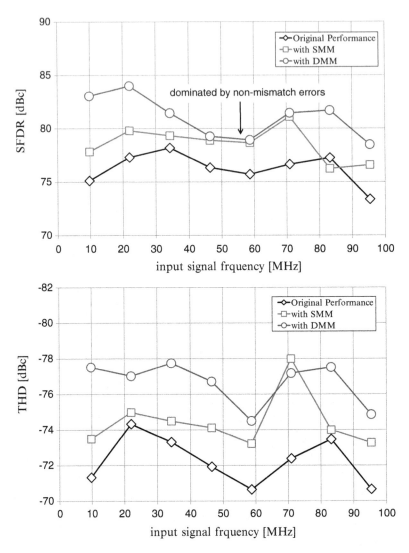

Fig. 7.13 Measured SFDR and THD at 200 MS/s

Compared to without mapping, the SFDR with DMM across the whole Nyquist band is improved from >73 dBc to >78 dBc and the THD (up to 11th harmonic) is improved from <-70 dBc to <-74 dBc, as shown in Fig. 7.13. The SMM shows much less improvement on SFDR and THD due to timing errors being not corrected. Though the SFDR with DMM is always better than the SFDR with SMM, compared to IM3 measurement results, the SFDR results do not show a clear differentiation in the improvement by correcting amplitude or timing errors. This is because the SFDR measurements check the spurs in the whole output spectrum, while the IM3

Table 7.2 DAC Performance summary with dynamic-mismatch mapping (DMM)

Technology	0.14 μm 1P6M 1.8 V CMOS baseline	
Resolution	14-bit	
Sampling rate	200 MHz	
Full-scale output	20 mA, 1 $V_{pp,diff}$ for dc signal	
(drive ability)	0.5 $V_{pp,diff}$ for ac signal (-2 dBm on 50Ω load)	
	without DMM	with DMM
max. INL/DNL	3.2 LSB/2 LSB	1.8 LSB/2 LSB
IM3, across whole Nyquist band	< -77 dBc	< -84 dBc
SFDR, across whole Nyquist band	>73 dBc	>78 dBc
THD, across whole Nyquist band	< -70 dBc	< -75 dBc
NSD, across whole Nyquist band	< -163 dBm/Hz, independent of mapping	
Power@200 MS/s	270 mW@1 V/1.8 V digital/analog supply	
Active area	2.4 mm^2	

measurements only check the third intermodulation component. In this design, the SFDR is affected by the switching interference caused by digital circuits. Since the mapping engine has four time-interleaved decoding sub-blocks operating at a quarter of the sampling frequency ($0.25 f_s = 50$ MHz), most of the switching interference is within the Nyquist band. Due to the low-ohmic substrate and no deep-nwell as shielding, switching interference from the mapping engine is coupled to and modulated by the DAC's output. During measurements, it is observed that the interference at the DAC output increases as the digital supply increases. This problem can be reduced by using deep-nwells to suppress the substrate coupling, implementing the mapping engine as current-mode logic or making the mapping engine not time-interleaved. Without modifying the design, using deep-nwell is recommended. With a deep-nwell, a 35 dB substrate noise suppression at 100 MHz is observed in [69].

An example of the DAC output spectrum measured at 200 MS/s and generating a single output tone at 95.4 MHz is shown in Figs. 7.14 and 7.15. As seen, the SFDR is 73.1 dBc without mapping. With traditional SMM to correct amplitude errors, the SFDR is improved by 1.5 dB to 74.6 dBc. With the proposed DMM to correct both amplitude and timing errors, the SFDR is improved by 5.7 dB to 78.8 dBc. The performance of this DMM DAC is summarized in Table 7.2.

7.3.3   Benchmark

A comparison of the SFDR with published state-of-the-art CMOS DACs at similar sampling rate (f_s) is shown in Fig. 7.16. As seen, compared to the DACs with conventional calibrations [3, 14], this work achieves much better SFDR and maintains it above 78 dBc in the whole 100 MHz Nyquist band. Compared to the best published DEM DAC [23], this work has 21 dB better NSD and comparable SFDR.

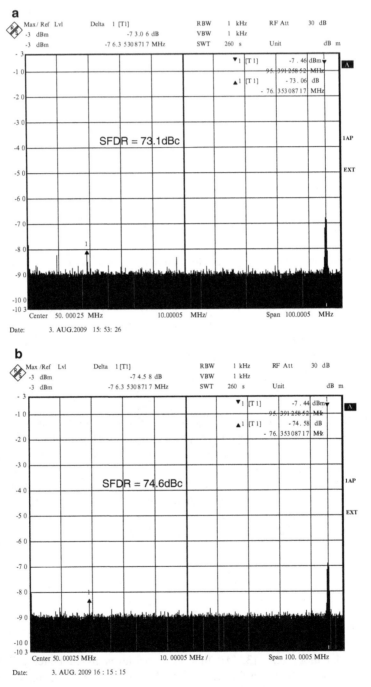

Fig. 7.14 DAC output spectrum with $f_i = 95.4\,\text{MHz}$ @200 MS/s (**a**) DAC output spectrum without mapping with $f_i = 95.4\,\text{MHz}$ @200 MS/s (**b**) DAC output spectrum with traditional SMM with $f_i = 95.4\,\text{MHz}$ @200 MS/s

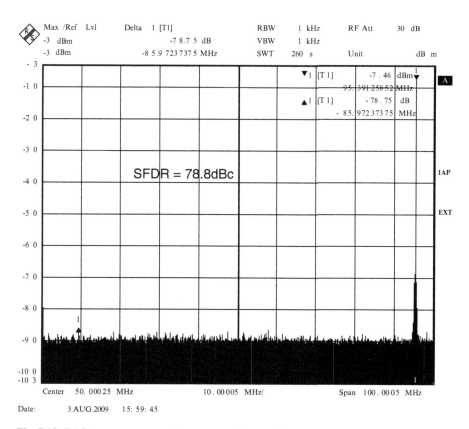

Fig. 7.15 DAC output spectrum with proposed DMM at $f_i = 95.4$ MHz @200 MS/s

A benchmark comparison with literature in the recent twelve years is given in Table 7.3. In Fig. 7.17, the comparison of the SFDR at very low signal frequencies (near DC) versus static effective number of bits (static ENOB, based on the INL) is given. Since the SFDR at very low signal frequency is dominated by amplitude errors, a 6 dB per effective bit trend line for the SFDR caused by amplitude errors is also plotted. As seen, this work achieves a medium static performance of a 12.2 bit ENOB. However, as concluded in Chap. 3, a good static performance does not mean a good dynamic performance, especially at high frequencies. At those high frequencies, other error sources, such as timing errors, finite output impedance and switching interference are dominant in the DAC's dynamic performance.

A comparison of the SFDR at high signal frequencies (near Nyquist frequency, i.e. $\approx 0.5 f_s$, unless specified in the table) are plotted in Fig. 7.18. As the proposed dynamic-mismatch mapping corrects both amplitude and timing errors, this work achieves a state-of-the-art dynamic performance up to 100 MHz which is a typical frequency range dominated by mismatch errors [70]. Compared to the DEM DAC in [23], this work [71, 72] has a competitive SFDR and has a 21 dB better NSD. For very high signal frequencies, finite output impedance dominates

Table 7.3 Benchmarking

Ref.	Year	Bit	INL/DNL [LSB]	f_s [MS/s]	SFDR [dBc] @low f_i	SFDR [dBc] @high f_i	Technology	Power	Design techniques
[2]	ISSCC'09	12	0.5/0.3	2,900	74	60[a]	65 nm CMOS, 2.5 V	188 mW	Always-on cascoding
[3]	ISSCC'07	13	0.8/0.4	200	83.7	54.5	0.13 um CMOS, 1.5 V	25 mW	Dynamic elelment matching (DEM)
[4]	ISSCC'06	14	–	100	74.4	77.8	0.18 um CMOS, 1.8 V	150 mW	DEM
[5]	ISSCC'06	6	–	20,000	–	50[b]	0.18 um SiGe, 1.8 V	360 mW	–
[6]	ISSCC'06	9	1/0.5	2	–	–	0.5 um CMOS, 5 V	0.3 mW	–
[7]	ISSCC'05	15	8	1,200	72	63	0.35 um BiCMOS, 3.3 V	6 W	DEM
[8]	ISSCC'05	12	–	1,600	62	55	GaAs, 5 V	1.2 W	RZ
[9]	ISSCC'05	12	–	1,700	64	50	0.35 um BiCMOS, 3 V	3 W	Current source array optimization
[10]	ISSCC'05	12	1/0.6	500	78	58	0.18 um CMOS, 1.8 V	216 mW	Current source array optimization
[11]	ISSCC'05	6	0.9/0.5	22,000	–	–	0.13 um BiCMOS, 3.3 V	1.2 W	–
[12]	ISSCC'04	14	1.8/0.8	1,400	–	60	0.18 um CMOS, 1.8 V	400 mW	Constant switching, crossover adjust
[13]	ISSCC'04	10	0.1/0.1	250	74	60	0.18 um CMOS, 1.8 V	4 mW	Fast switching+dummy
[14]	ISSCC'04	14	0.65/0.55	200	85	44	0.18 um CMOS, 1.8 V	97 mW	Current source calibration
[15]	ISSCC'03	16	1/0.25	400	95	73	0.25 um CMOS, 3.3 V	400 mW	Current source calibration, bootstrap
[16]	ISSCC'03	14	0.43/0.34	100	82	62	0.13 um CMOS, 1.5 V	16.7 mW	Foreground current source calibration
[17]	ISSCC'01	12	0.3/0.25	500	75	35	0.35 um CMOS, 3 V	110 mW	Current source array optimization
[18]	ISSCC'00	14	0.5/0.5	100	82	72	0.35 um CMOS, 3.3 V	180 mW	Current source calibration + RZ
[19]	ISSCC'99	14	0.3/0.2	150	84	50	0.5 um CMOS, 2.7 V	300 mW	Q^2 current source array optimization
[20]	ISSCC'99	14	0.5/0.5	60	85	75	0.8 um CMOS, 5 V	750 mW	RZ
[21]	ISSCC'98	10	0.2/0.1	250	71	57	0.5 um CMOS, 5 V	100 mW	Local biasing
[22]	ISSCC'98	12	0.6/0.3	300	70	40	0.5 um CMOS, 3.3 V	320 mW	Cascode switches

[23]	VLSI'07	14	3.5/1	150	83	83	0.18 um CMOS, 1.8 V	127 mW	DEM
[24]	ESSCIRC'06	8	0.25/0.25	600	68	–	0.13 um CMOS, 1.2 V	2.4 mW	–
[25]	ESSCIRC'05	12	0.4/0.6	50	80	60	0.25 um CMOS, 3.3 V	270 mW	Current source calibration
[26]	ESSCIRC'04	14	0.7/0.45	130	80	40	0.25 um CMOS, 3.3 V	103 mW	Static-mismatch mapping (SMM)
[27]	JSSC'06	5	–	32,000	31	30	300 GHz ft Bipolar	4.4 W	Sine-weighted DAC
[28]	JSSC'06	12	0.38/0.44	180	72	62	0.25 um CMOS, 3.3 V	155 mW	–
[29]	JSSC'03	12	0.4/0.3	320	95	45	0.18 um CMOS, 1.8 V	60 mW	DEM
[30]	JSSC'03	14	0.3/0.3	300	72	68	0.25 um CMOS, 3.3 V	53 mW	Current source calibration + RZ
[31]	JSSC'01	10	0.2/0.15	1,000	72	61	0.35 um CMOS, 3 V	110 mW	Boosted output impedance
[32]	JSSC'98	12	0.6/0.3	300	70	40	0.5 um CMOS, 3.3 V	320 mW	Optimized switch & latch
This work	VLSI'10	14	1.8/2	200	83	78	0.14 um CMOS, 1.8 V	270 mW	Dynamic-mismatch mapping (DMM)

[a] @ 550 MHz
[b] @ 186 MHz

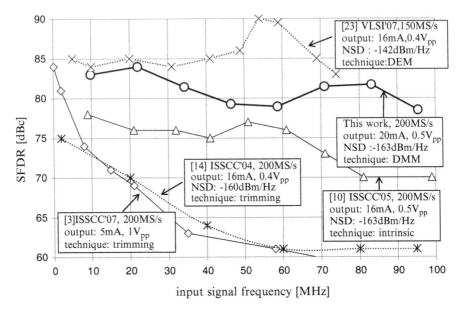

Fig. 7.16 SFDR comparison with state-of-the-art CMOS DACs at similar f_s

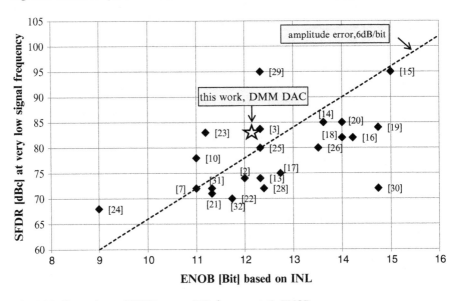

Fig. 7.17 Comparison of SFDR at near-DC f_i versus static ENOB

the performance [2]. In Fig. 7.18, the trend lines for timing errors and finite output impedance (Z_{out}) are also plotted to empirically show the trend of performance limitation. Since the normalized input signal frequency is always about 0.5 in

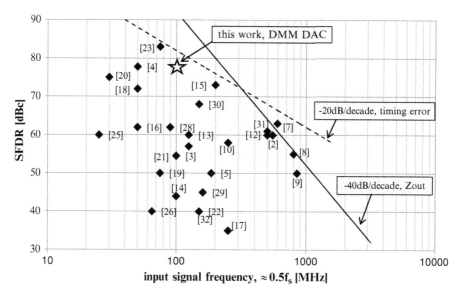

Fig. 7.18 Comparison of SFDR at near-Nyquist f_i

this plot, the trend of the SFDR with signal frequencies for timing errors is 20 dB/decade. Note that these two lines do not show the performance limitation quantitatively, but only the trends.

7.4 Conclusions

In this chapter, a 14-bit current-steering DAC with the proposed digital calibration technique, i.e. dynamic-mismatch mapping (DMM), is demonstrated. The intrinsic DAC core has a performance of SFDR>65 dBc across the whole Nyquist band at 650 MS/s. Silicon experimental results show that the proposed dynamic-mismatch mapping (DMM) can correct both amplitude and timing errors, without increasing the noise floor. With DMM, the DAC achieves a state-of-the-art performance of SFDR>78 dBc, IM3<-83 dBc and NSD<-163 dBm/Hz across the whole Nyquist band at 200 MS/s. Comparing to the intrinsic performance shows at least 5 dB linearity improvement in the whole Nyquist band by DMM.

Regarding the comparison to other digital calibration techniques: compared to static-mismatch mapping (SMM), DMM can provide performance improvement in the whole Nyquist band by correcting both amplitude and timing errors; compared to dynamic element matching (DEM), DMM does not increase the noise floor.

Benchmark comparing to other published DACs shows that both the intrinsic DAC core and the DMM DAC have a state-of-the-art performance, especially for the dynamic performance.

Chapter 8
Conclusions

In the signal frequency range from DC to several hundreds of MHz, mismatch errors, including amplitude and timing errors, are typical dominant factors in the linearity of current-steering DACs. Moreover, as signal and sampling frequencies increase, the effect of timing errors becomes more and more dominant over that of amplitude errors.

New parameters, i.e. dynamic-INL and dynamic-DNL, can efficiently evaluate the dynamic-matching performance of current cells. Compared to traditional static INL and DNL, dynamic-INL and dynamic-DNL describe the matching performance between current cells more completely and accurately. The frequency-dependent characteristic of dynamic-INL and dynamic-DNL allows to balance the weight between amplitude and timing errors to achieve the best performance for different applications.

Technology limitations make it very difficult for an intrinsic DAC to achieve high performance at high sampling frequencies. A smart DAC can potentially achieve a better performance than an intrinsic DAC since it measures and corrects the actual error information.

A novel digital calibration technique called dynamic-mismatch mapping (DMM) for smart DACs can significantly improve both the static and dynamic performance of current-steering DACs, without increasing the noise floor. The proposed DMM reduces the dynamic-INL by optimizing the switching sequence of thermometer current cells, so that the effect of both amplitude and timing errors can be corrected. Compared to static-mismatch mapping (SMM), DMM can improve the performance across the whole Nyquist band, especially at high frequencies. This advantage of DMM over SMM is due to the fact that the effects of both amplitude and timing errors are corrected. Compared to dynamic element matching (DEM), DMM does not increase the noise floor because the mismatch effect is reduced instead of randomized. This technique is validated by both theoretical and silicon measurement results.

A feature of the proposed DMM is *analog measurement, digital actuation*, i.e. the error is measured in an analog way and corrected in a digital way. Therefore,

Y. Tang et al., *Dynamic-Mismatch Mapping for Digitally-Assisted DACs*, Analog Circuits 157
and Signal Processing 92, DOI 10.1007/978-1-4614-1250-2_8,
© Springer Science+Business Media New York 2013

it can be easily stacked on other techniques with analog actuation, e.g. current source calibration. In theory, as long as the performance is dominated by mismatch errors, DMM always provides almost 10 dB improvement in the whole Nyquist band, without increasing the noise floor and regardless the starting point is 70 dB or 90 dB. This makes the proposed DMM a very attractive "last-mile" solution on the performance pyramid.

The proposed on-chip dynamic-mismatch sensor based on a zero-IF receiver can accurately measure the dynamic-mismatch errors of current cells.

The design example of a 14-bit 0.14 μm current-steering CMOS DAC which can be switched between intrinsic-DAC and smart-DAC modes has been demonstrated with experimental results. In the intrinsic-DAC mode, the 14-bit 650 MS/s intrinsic DAC core achieves a performance of SFDR>65 dBc across the whole 325 MHz Nyquist band. In the smart-DAC mode, compared to the intrinsic DAC performance, DMM provides at least 5 dB linearity improvement across the whole Nyquist band at 200 MS/s, without increasing the noise floor. The 14-bit 200 MS/s smart DAC with DMM achieves a performance of SFDR>78 dBc, IM3<-83 dBc and NSD<-163 dBm/Hz across the whole Nyquist band. Both of two modes achieve a state-of-the-art performance.

References

1. W. Kester, in *Data Conversion Handbook (Analog Devices)*. Newnes, 2004
2. C.-H. Lin, F. van der Goes, J. Westra, J. Mulder, Y. Lin, E. Arslan, E. Ayranci, X. Liu, K. Bult, A 12b 2.9 GS/s DAC with IM3<-60 dBc beyond 1 GHz in 65 nm CMOS, in *Solid-State Circuits Conference – Digest of Technical Papers, 2009. ISSCC 2009. IEEE International* (2009), pp. 74–75, 75a
3. M. Clara, W. Klatzer, B. Seger, A. Di Giandomenico, L. Gori, A 1.5 V 200 MS/s 13b 25 mW DAC with randomized nested background calibration in 0.13 μm CMOS, in *Solid-State Circuits Conference, 2007. ISSCC 2007. Digest of Technical Papers. IEEE International* (2007), pp. 250–600
4. K.L. Chan, I. Galton, A 14b 100 MS/s DAC with fully segmented dynamic element matching, in *Solid-State Circuits Conference, 2006. ISSCC 2006. Digest of Technical Papers. IEEE International* (2006), pp. 2390–2399
5. D. Baranauskas, D. Zelenin, A 0.36 W 6b up to 20 GS/s DAC for UWB wave formation, in *Solid-State Circuits Conference, 2006. ISSCC 2006. Digest of Technical Papers. IEEE International* (2006), pp. 2380–2389
6. I. Knausz, R. Bowman, A 250 μW 0.042 mm^2 2 MS/s 9b DAC for liquid crystal display drivers, in *Solid-State Circuits Conference, 2006. ISSCC 2006. Digest of Technical Papers. IEEE International* (2006), pp. 599–608
7. B. Jewett, J. Liu, K. Poulton, A 1.2 GS/s 15b DAC for precision signal generation, in *Solid-State Circuits Conference, 2005. Digest of Technical Papers. ISSCC. 2005 IEEE International*, vol. 1 (2005), pp. 110–587
8. M.-J. Choe, K.-H. Baek, M. Teshome, A 1.6 GS/s 12b return-to-zero GaAs RF DAC for multiple Nyquist operation, in *Solid-State Circuits Conference, 2005. Digest of Technical Papers. ISSCC. 2005 IEEE International*, vol. 1 (2005), pp. 112–587
9. K.-H. Baek, E. Merlo, M.-J. Choe, A. Yen, M. Sahrling, A 1.7 GHz 3 V direct digital frequency synthesizer with an on-chip DAC in 0.35 μm SiGe BiCMOS, in *Solid-State Circuits Conference, 2005. Digest of Technical Papers. ISSCC. 2005 IEEE International*, vol. 1 (2005), pp. 114–587
10. K. Doris, J. Briaire, D. Leenaerts, M. Vertreg, A. van Roermund, A 12b 500 MS/s DAC with >70 dB SFDR up to 120 MHz in 0.18 μm CMOS, in *Solid-State Circuits Conference, 2005. Digest of Technical Papers. ISSCC. 2005 IEEE International*, vol. 1 (2005), pp. 116–588
11. P. Schvan, D. Pollex, T. Bellingrath, A 22 GS/s 6b DAC with integrated digital ramp generator, in *Solid-State Circuits Conference, 2005. Digest of Technical Papers. ISSCC. 2005 IEEE International*, vol. 1 (2005), pp. 122–588

Y. Tang et al., *Dynamic-Mismatch Mapping for Digitally-Assisted DACs*, Analog Circuits and Signal Processing 92, DOI 10.1007/978-1-4614-1250-2,
© Springer Science+Business Media New York 2013

12. B. Schafferer, R. Adams, A 3 V CMOS 400 mW 14b 1.4 GS/s DAC for multi-carrier applications, in *Solid-State Circuits Conference, 2004. Digest of Technical Papers. ISSCC. 2004 IEEE International*, vol. 1 (2004), pp. 360–532

13. J. Deveugele, M. Steyaert, A 10b 250 MS/s binary-weighted current-steering DAC, in *Solid-State Circuits Conference, 2004. Digest of Technical Papers. ISSCC. 2004 IEEE International*, vol. 1 (2004), pp. 362–532

14. Q. Huang, P. Francese, C. Martelli, J. Nielsen, A 200 MS/s 14b 97 mW DAC in 0.18 μm CMOS, in *Solid-State Circuits Conference, 2004. Digest of Technical Papers. ISSCC. 2004 IEEE International*, vol. 1 (2004), pp. 364–532

15. W. Schofield, D. Mercer, L. Onge, A 16b 400 MS/s DAC with <-80 dBc IMD to 300 MHz and <-160 dBm/Hz noise power spectral density, in *Solid-State Circuits Conference, 2003. Digest of Technical Papers. ISSCC. 2003 IEEE International*, vol. 1 (2003), pp. 126–482

16. Y. Cong, R. Geiger, A 1.5-V 14-bit 100-MS/s self-calibrated DAC. IEEE J. Solid-State Circ. **38**(12), 2051–2060 (2003)

17. A. Van Den Bosch, M. Borremans, M. Steyaert, W. Sansen, A 12b 500 MSample/s current-steering CMOS D/A converter, in *Solid-State Circuits Conference, 2001. Digest of Technical Papers. ISSCC. 2001 IEEE International* (2001), pp. 366–367, 466

18. A. Bugeja, B.-S. Song, A self-trimming 14-b 100-MS/s CMOS DAC. IEEE J. Solid-State Circ. **35**(12), 1841–1852 (2000)

19. J. Vandenbussche, G. Van der Plas, A. Van den Bosch, W. Daems, G. Gielen, M. Steyaert, W. Sansen, A 14b 150 Msample/s update rate Q^2 random walk CMOS DAC, in *Solid-State Circuits Conference, 1999. Digest of Technical Papers. ISSCC. 1999 IEEE International* (1999), pp. 146–147

20. A. Bugeja, B.-S. Song, P. Rakers, S. Gillig, A 14b 100 Msample/s CMOS DAC designed for spectral performance, in *Solid-State Circuits Conference, 1999. Digest of Technical Papers. ISSCC. 1999 IEEE International* (1999), pp. 148–149

21. C.-H. Lin, K. Bult, A 10-b 500-MSample/s CMOS DAC in 0.6 mm². IEEE J. Solid-State Circ. **33**(12), 1948–1958 (1998)

22. A. Marques, J. Bastos, A. Van den Bosch, J. Vandenbussche, M. Steyaert, W. Sansen, A 12b accuracy 300 Msample/s update rate CMOS DAC, in *Solid-State Circuits Conference, 1998. Digest of Technical Papers. 1998 IEEE International* (1998), pp. 216–217, 440

23. K.L. Chan, J. Zhu, I. Galton, A 150 MS/s 14-bit segmented DEM DAC with greater than 83 dB of SFDR across the Nyquilst band, in *2007 IEEE Symposium on VLSI Circuits* (2007), pp. 200–201

24. N. Ghittori, A. Vigna, P. Malcovati, S. D'Amico, A. Baschirotto, A 1.2-V, 600-MS/s, 2.4-mW DAC for WLAN 802.11 and 802.16 Wireless Transmitters, in *Solid-State Circuits Conference, 2006. ESSCIRC 2006. Proceedings of the 32nd European* (2006), pp. 404–407

25. G. Radulov, P. Quinn, H. Hegt, A. van Roermund, An on-chip self-calibration method for current mismatch in D/A converters, in *Solid-State Circuits Conference, 2005. ESSCIRC 2005. Proceedings of the 31st European* (2005), pp. 169–172

26. T. Chen, P. Geens, G. Van der Plas, W. Dehaene, G. Gielen, A 14-bit 130-MHz CMOS current-steering DAC with adjustable INL, in *Solid-State Circuits Conference, 2004. ESSCIRC 2004. Proceeding of the 30th European* (2004), pp. 167–170

27. S. Turner, D. Kotecki, Direct digital synthesizer with sine-weighted DAC at 32-GHz clock frequency in InP DHBT technology. IEEE J. Solid-State Circ. **41**(10), 2284–2290 (2006)

28. K. Gulati, M. Peng, A. Pulincherry, C. Munoz, M. Lugin, A. Bugeja, J. Li, A. Chandrakasan, A highly integrated CMOS analog baseband transceiver with 180MSPS 13-bit Pipelined CMOS ADC and Dual 12-bit DACs. IEEE J. Solid-State Circ. **41**(8), 1856–1866 (2006)

29. K. O'Sullivan, C. Gorman, M. Hennessy, V. Callaghan, A 12-bit 320-MSample/s current-steering CMOS D/A converter in 0.44 mm². IEEE J. Solid-State Circ. **39**(7), 1064–1072 (2004)

30. J. Hyde, T. Humes, C. Diorio, M. Thomas, M. Figueroa, A 300-MS/s 14-bit digital-to-analog converter in logic CMOS. IEEE J. Solid-State Circ. **38**(5), 734–740 (2003)

31. A. van den Bosch, M. Borremans, M. Steyaert, W. Sansen, A 10-bit 1-GSample/s Nyquist current-steering CMOS D/A converter. IEEE J. Solid-State Circ. **36**(3), 315–324 (2001)

32. J. Bastos, A. Marques, M. Steyaert, W. Sansen, A 12-bit intrinsic accuracy high-speed CMOS DAC. IEEE J. Solid-State Circ. **33**(12), 1959–1969 (1998)
33. K. Lakshmikumar, R. Hadaway, M. Copeland, Characterisation and modeling of mismatch in MOS transistors for precision analog design. IEEE J. Solid-State Circ. **21**(6), 1057–1066 (1986)
34. M. Pelgrom, A. Duinmaijer, A. Welbers, Matching properties of MOS transistors. IEEE J. Solid-State Circ. **24**(5), 1433–1439 (1989)
35. C. Conroy, W. Lane, M. Moran, Statistical design techniques for D/A converters. IEEE J. Solid-State Circ. **24**(4), 1118–1128 (1989)
36. A. van den Bosch, M. Steyaert, W. Sansen, *Static and Dynamic Performance Limitations for High Speed D/A Converters* (Kluwer, Dordecht, 2004)
37. G.I. Radulov, M. Heydenreich, R.W. van der Hofstad, J.A. Hegt, A.H.M. van Roermund, Brownian-bridge-based statistical analysis of the DAC INL caused by current mismatch. IEEE Trans. Circ. Syst. II: Express Briefs **54**(2), 146–150 (2007)
38. Y. Tang, H. Hegt, A. van Roermund, K. Doris, J. Briaire, Statistical analysis of mapping technique for timing error correction in current-steering dacs, in *IEEE International Symposium on Circuits and Systems, 2007. ISCAS 2007* (2007), pp. 1225–1228
39. J. Wikner, Studies on CMOS digital-to-analog converters, PhD Thesis, Linkoping University, 2001
40. K. Doris, A. van Roermund, D. Leenaerts, *Wide-Bandwidth High Dynamic Range D/A Converters* (Springer, Berlin, 2006)
41. T. Chen, G. Gielen, The analysis and improvement of a current-steering DACs dynamic SFDR -I: the cell-dependent delay differences. IEEE Trans. Circ. Syst. I: Regular Papers **53**(1), 3–15 (2006)
42. R. Lyons, A differentiator with a difference @ONLINE. [Online] (2007). Jan 10, 2010 Available: http://www.dsprelated.com/showarticle/35.php
43. J. Gonzalez, E. Alarcon, Clock-jitter induced distortion in high speed CMOS switched-current segmented digital-to-analog converters, in *The 2001 IEEE International Symposium on Circuits and Systems, 2001. ISCAS 2001*, vol. 1 (2001), pp. 512–515
44. K. Doris, A. van Roermund, D. Leenaerts, A general analysis on the timing jitter in D/A converters, in *IEEE International Symposium on Circuits and Systems, 2002. ISCAS 2002*, vol. 1 (2002), pp. I–117–I–120
45. R. van der Plassche, *Integrated Analog-to-Digital and Digital-to-Analog Converters* (Kluwer, Dordecht, 1994)
46. P. Palmers, M. Steyaert, A 11 mW 68 dB SFDR 100 MHz bandwidth sigma-delta DAC based on a 5-bit 1 GS/s core in 130 nm, in *34th European Solid-State Circuits Conference, 2008. ESSCIRC 2008* (2008), pp. 214–217
47. B. Razavi, *RF Microelectronics* (Prentice Hall, Englewood Cliffs, 1997)
48. H.J. Bergveld, E.C. Dijkmans, H.A.H. Termeer, "Current DAC design and measurements," *Philips/NXP Technical Note*, 2001
49. S. Park, G. Kim, S.-C. Park, W. Kim, A digital-to-analog converter based on differential-quad switching. IEEE J. Solid-State Circ. **37**(10), 1335–1338 (2002)
50. G. Radulov, P. Quinn, P. Harpe, H. Hegt, A. van Roermund, Parallel current-steering D/A converters for flexibility and smartness, in *IEEE International Symposium on Circuits and Systems, 2007. ISCAS 2007* (2007), pp. 1465–1468
51. G. Radulov, Flexible and self-calibrating current-steering digital-to-analog converters: analysis, classification and design, PhD Thesis, Eindhoven University of Technology, 2010
52. F.F. Dai, W. Ni, S. Yin, R. Jaeger, A direct digital frequency synthesizer with fourth-order phase domain $\Sigma\Delta$ noise shaper and 12-bit current-steering DAC. IEEE J. Solid-State Circ. **41**(4), 839–850 (2006)
53. K.L. Chan, N. Rakuljic, I. Galton, Segmented dynamic element matching for high-resolution digital-to-analog conversion. IEEE Trans. Circ. Syst. I: Regular Papers **55**(11), 3383–3392 (2008)

54. B. Catteau, P. Rombouts, L. Weyten, A digital calibration technique for the correction of glitches in high-speed DACs, in *IEEE International Symposium on Circuits and Systems, 2007. ISCAS 2007* (2007), pp. 1477–1480
55. Y. Tang, H. Hegt, A. van Roermund, Predictive timing error calibration technique for RF current-steering DACs, in *IEEE International Symposium on Circuits and Systems, 2008. ISCAS 2008* (2008), pp. 228–231
56. K. Rafeeque, V. Vasudevan, A new technique for on-chip error estimation and reconfiguration of current-steering digital-to-analog converters. IEEE Trans. Circ. Syst. I: Regular Papers **52**(11), 2348–2357 (2005)
57. T. Chen, G. Gielen, The analysis and improvement of a current-steering DAC's dynamic SFDR -II: the output-dependent delay differences. IEEE Trans. Circ. Syst. I: Regular Papers **54**(2), 268–279 (2007)
58. T. Chen, G. Gielen, A 14-bit 200-MHz current-steering DAC with switching-sequence post-adjustment calibration. IEEE J. Solid-State Circ. **42**(11), 2386–2394 (2007)
59. K. Doris, C. Lin, D. Leenaerts, A. van Roermund, D/a conversion: amplitude and time error mapping optimization, in *The 8th IEEE International Conference on Electronics, Circuits and Systems, 2001. ICECS 2001*, vol. 2 (2001), pp. 863–866
60. Y. Tang, H. Hegt, A. van Roermund, Ddl-based calibration techniques for timing errors in current-steering dacs, in *2006 IEEE International Symposium on Circuits and Systems, 2006. ISCAS 2006. Proceedings* (2006), p. 4
61. K. Doris, A. van Roermund, C. Lin, D. Leenaerts, Error optimization in digital to analog conversion. Patent WO/2003/021790, 2003
62. M. Terrovitis, R. Meyer, Noise in current-commutating CMOS mixers. IEEE J. Solid-State Circ. **34**(6), 772–783 (1999)
63. H. Darabi, A. Abidi, Noise in RF-CMOS mixers: a simple physical model. IEEE J. Solid-State Circ. **35**(1), 15–25 (2000)
64. W. Redman-White, D. Leenaerts, 1/f noise in passive CMOS mixers for low and zero IF integrated receivers, in *Solid-State Circuits Conference, 2001. ESSCIRC 2001. Proceedings of the 27th European* (2001), pp. 41–44
65. E. Sacchi, I. Bietti, S. Erba, L. Tee, P. Vilmercati, R. Castello, A 15 mW, 70 kHz 1/f corner direct conversion CMOS receiver, in *Custom Integrated Circuits Conference, 2003. Proceedings of the IEEE 2003* (2003), pp. 459–462
66. M. Valla, G. Montagna, R. Castello, R. Tonietto, I. Bietti, A 72-mW CMOS 802.11a direct conversion front-end with 3.5-dB NF and 200-kHz 1/f noise corner. IEEE J. Solid-State Circ. **40**(4), 970–977 (2005)
67. R. van Veldhoven, A tri-mode continuous-time sigma-delta modulator with switched-capacitor feedback DAC for a GSM-EDGE/CDMA2000/UMTS receiver, in *Solid-State Circuits Conference, 2003. Digest of Technical Papers. ISSCC. 2003 IEEE International*, vol. 1 (2003), pp. 60–477
68. R. van Veldhoven, A triple-mode continuous-time Sigma-Delta modulator with switched-capacitor feedback DAC for a GSM-EDGE/CDMA2000/UMTS receiver. IEEE J. Solid-State Circ. **38**(12), 2069–2076 (2003)
69. K.W. Chew, J. Zhang, K. Shao, W.B. Loh, S.-F. Chu, Impact of deep N-well implantation on substrate noise coupling and RF transistor performance for systems-on-a-chip integration, in *Solid-State Device Research Conference, 2002. Proceeding of the 32nd European* (2002), pp. 251–254
70. Y. Tang, H. Hegt, A. van Roermund, Smart DACs: on the road towards Giga-Hz RF DACs, in *Proceedings ProRisc 2007, Veldhoven, the Netherlands* (2007)
71. Y. Tang, J. Briaire, K. Doris, R. van Veldhoven, P. van Beek, H. Hegt, A. van Roermund, A 14 bit 200 MS/s DAC With SFDR >78 dBc, IM3 < -83 dBc and NSD <-163 dBm/Hz across the whole Nyquist band enabled by dynamic-mismatch mapping, in *2010 IEEE Symposium on VLSI Circuits (VLSIC)* (2010), pp. 151–152
72. Y. Tang, J. Briaire, K. Doris, R. van Veldhoven, P. van Beek, H. Hegt, A. van Roermund, A 14 bit 200 MS/s DAC with SFDR >78 dBc, IM3 < -83 dBc and NSD <-163 dBm/Hz across the whole Nyquist band enabled by dynamic-mismatch mapping. IEEE J. Solid-State Circ. **46**(6), 1371–1381 (2011)

Index

A
ACLR, 12
ADC, 7, 55, 134
Amplitude error, 24, 103

C
Calibration, 70, 79–81, 83
Capacitive DAC, 15
Cascode, 73
Common duty-cycle error, 56, 73
Constant switching, 75
Correlation, 28, 40
Current source, 24
Current-mode logic (CML), 138, 145
Current-Steering DAC (CS-DAC), 16

D
DA architecture
    Binary architecture, 12
    Segmented architecture, 14
    Unary architecture, 13
DA conversion
    Frequency domain response, 7
    Time domain response, 5
DAC, 5
DC bleeding, 122
Delay error, 34
Demodulation, 98, 100, 126
Differentiator, 37
Digital Calibration, 81
DNL, 10, 26, 50
Duty-cycle error, 34, 56
Dynamic element matching (DEM), 78, 79, 115

Dynamic mismatch, 32, 46
Dynamic-DNL, 47, 50, 106
Dynamic-INL, 47, 50, 93, 106
Dynamic-mismatch mapping (DMM), 85, 91, 105, 143

E
ENOB, 17

F
Filter, 124
Finite output impedance, 61, 73
Flicker (1/f) noise, 122, 128, 129

G
Gain errors, 8
Gain-boosting, 125

H
Harmonic suppression, 77

I
IM3, 146
INL, 9, 26, 50
Interference, 64, 75
Intermodulation Distortion (IMD), 11
Intrinsic DAC, 69, 86, 138

J
Jitter, 51, 71

L
Layout techniques, 78, 139
LVDS, 139

M
Mapping, 83, 85
Mapping engine, 143
Measurement frequency, 101, 103, 117, 120,
    126
Mismatch, 23, 32, 77
Mixer, 122, 129, 133

N
Non-overlapping, 133
Non-return-to-zero (NRZ), 6
NSD, 11, 146
Nyquist, 7, 8

O
Offset, 8, 119
OTA, 124, 129
Oversampling, 8

P
Performance pyramid, 116
Phase detector, 81

Q
Quantization noise, 7

R
Resistive DAC, 14

R
Return-to-zero (RZ), 6
Robustness, 113

S
Sampling frequency, 7
Sampling period, 6
SDR, 28, 41
Sensor, 119, 120, 125, 143
SFDR, 11, 17, 26, 31, 36, 43, 106, 142, 148
Sinc attenuation, 7
Smart DAC, 69, 79, 80, 86, 116, 143
SNDR, 11
SNR, 11
Sorting, 143
State of The Art, 17, 142, 151
Static mismatch, 23
Static-mismatch mapping (SMM), 91, 115,
    148
Switched-capacitor, 131
Switching sequence, 83, 91, 94, 105, 113

T
Taylor expansion, 63
THD, 11, 26, 31, 36, 43, 106, 148
Thermal noise, 129
Time-interleave, 145
Timing error, 33, 103
Trans-impedance amplifier (TIA), 122, 124
Transfer function, 47, 125

Z
Zero-IF receiver, 119
Zero-order hold (ZOH), 7